T0299339

Lattice Path Combinatorics and Special Counting Sequences

This book endeavors to deepen our understanding of lattice path combinatorics, explore key types of special sequences, elucidate their interconnections, and concurrently champion the author's interpretation of the "combinatorial spirit".

The author intends to give an up-to-date introduction to the theory of lattice path combinatorics, its relation to those special counting sequences important in modern combinatorial studies, such as the Catalan, Schröder, Motzkin, Delannoy numbers, and their generalized versions. Brief discussions of applications of lattice path combinatorics to symmetric functions and connections to the theory of tableaux are also included. Meanwhile, the author also presents an interpretation of the "combinatorial spirit" (i.e., "counting without counting", bijective proofs, and understanding combinatorics from combinatorial structures internally, and more), hoping to shape the development of contemporary combinatorics.

Lattice Path Combinatorics and Special Counting Sequences: From an Enumerative Perspective will appeal to graduate students and advanced undergraduates studying combinatorics, discrete mathematics, or computer science.

Chunwei Song, a combinatorialist and graph theorist, is a professor of mathematics at Peking University. He received his PhD from the University of Pennsylvania in 2004. He also held faculty positions at Boston College, MA and the Tokyo Institute of Technology, Japan. In 2010, he was a visiting associate professor at the University of Delaware. At Peking University, seven students have received PhD degrees under his supervision, specializing in the fields of lattice path combinatorics, extremal combinatorics, or mathematical philosophy.

Lattice Path Combinatorics and Special Counting Sequences

From an Enumerative Perspective

Chunwei Song

CRC Press
Taylor & Francis Group
Boca Raton London New York

CRC Press is an imprint of the
Taylor & Francis Group, an **informa** business

Designed cover image: © Chunwei Song

First edition published 2025
by CRC Press
2385 NW Executive Center Drive, Suite 320, Boca Raton FL 33431

and by CRC Press
4 Park Square, Milton Park, Abingdon, Oxon, OX14 4RN

CRC Press is an imprint of Taylor & Francis Group, LLC

© 2025 Chunwei Song

ISBN: 978-1-032-67175-8 (hbk)
ISBN: 978-1-032-83557-0 (pbk)
ISBN: 978-1-003-50991-2 (ebk)

DOI: 10.1201/9781003509912

Typeset in Nibus font
by KnowledgeWorks Global Ltd.

To my teachers,
colleagues,
and students.

Contents

List of Tables

List of Figures

Preface

This book was rooted in my doctoral dissertation at the University of Pennsylvania. My first exposure to combinatorial mathematics occurred in 1999 at a combinatorics course given by Herbert S. Wilf. In subsequent years, I studied different subject matter involving combinatorics including graph theory with Jennifer Morse, Amy Myers, Paul Seymour, and Felix Lazebnik. But it was my thesis advisor James Haglund who first introduced to me this fascinating theme of *lattice path combinatorics* in a systematic manner. Influenced and indebted, I am grateful to all my teachers, for their guidance and encouragement.

During the fall semester of 2005, I was invited to teach a special graduate course on topics in advanced combinatorics at the Tokyo Institute of Technology. In preparation for the course, I created a pamphlet of unpublished lecture notes covering mainly *extremal combinatorics* and the *probabilistic method used in combinatorics*. For the past decade and longer, I taught multiple times a cross-listed combinatorics course at Peking University to both undergraduate and graduate students. Eventually, I wrote a textbook in Chinese together with my colleague, Prof. Rongquan Feng. A portion of this current text is an expansion of selected contents of my previous work.

I wish to thank all of the participants of my lectures. At Peking University, my teaching assistants include Bin Wang, Tongyuan Zhao, Feng Zhao, Han Hu, Suman Xia, Bowen Yao, Huabin Cao, Xinyang Ye, and Ningxin Zhang. So far five of them, Drs. Tongyuan, Feng, Han, Suman, and Bowen, obtained PhDs in the direction of lattice path combinatorics or related fields under my supervision. Likewise, the students in class were excellent some of whom provided valuable suggestions.

I appreciate the constant support of my colleagues at Peking University. It has been a great pleasure to discuss mathematics with my friends in China and in other places. To name a few from an otherwise neverending list, limited by space, I would mention Professors Bill Chen, Guo-Niu Han, Colton Magnant, Toufik Mansour, Yi Wang, Catherine Yan, Yeong-Nan Yeh, and Chuanming Zong.

The field of lattice path combinatorics has expanded substantially, making it impractical to provide a comprehensive discussion within the confines of this Preface. Therefore, this current work serves as a partial introduction. It is my aspiration that a future chapter dedicated to this topic in a forthcoming *Handbook of Combinatorics* will offer a more comprehensive treatment.

The intended readership of this book primarily comprises graduate students and accomplished undergraduate students in mathematics, tailored to individual circumstances. Additionally, a portion of the content serves as a reference for researchers, thus presented with less detailed elaboration.

Finally, I hope to express my gratitude to Peking University.

Chunwei Song
Beijing, P.R. China
March 3, 2024

CHAPTER 1

Introduction

1.1 OVERVIEW: THE SUBJECT AND SPIRIT

Over the past few decades, lattice path combinatorics has emerged as a vibrant and fertile subfield within the broader realm of combinatorics. Its allure lies in its inherent elegance, coupled with its myriad applications across diverse areas of mathematics. This captivating discipline is devoted to the enumeration of lattice paths across a range of configurations, offering not only aesthetic appeal but also profound insights into fundamental combinatorial structures. Moreover, Lattice Path Combinatorics frequently intersects with the realm of combinatorial statistics, forging robust connections that enrich both fields and foster interdisciplinary collaboration.

Furthermore, the theory of special sequences occupies a central position in the landscape of combinatorial mathematics. A seminal contribution to this domain is the meticulous compilation of 2,372 sequences of integers, sourced from a vast array of mathematical and scientific domains, curated by Sloane [Slo73]. This monumental effort continues to evolve, with ongoing contributions and updates meticulously chronicled at [Slo], where the database now boasts an impressive repository of over 370,000 entries. Many combinatorial paths find their enumeration intricately intertwined with celebrated integer sequences, including Dyck paths, Delannoy paths, Schröder paths, and their myriad variations and extensions. This symbiotic relationship underscores the indispensable role of special sequences in unraveling the complexities of lattice path configurations and advancing our understanding of combinatorial phenomena across disciplines.

DOI: 10.1201/9781003509912-1

A combinatorial statistic, denoted as stat, defined on a set A, serves as a mapping from elements of A to non-negative integers, each bearing significant combinatorial implications. This mapping assigns a numerical value, stat(a), to each element a in A, reflecting its relevance to a specific combinatorial property. Consider A as the ensemble of a particular class of lattice paths. By summing the expression $q^{\text{stat}(a)}$ over all elements a in A, we derive a polynomial in q. Remarkably, when q is substituted with 1, this polynomial yields the cardinality of A, symbolized as $|A|$. The beauty of this polynomial lies in its capacity to encapsulate various interpretations and convey valuable insights, contingent upon the definition of the combinatorial statistic stat. Depending on the specific context, this polynomial may unveil diverse combinatorial phenomena, offering a versatile tool for analysis and exploration within the realm of lattice path combinatorics.

In the exploration of lattice path combinatorics, several indispensable tools emerge as fundamental to understanding and advancing the field. Among these tools are decomposition techniques, which facilitate the dissection of complex problems into more manageable components, enabling a more nuanced analysis. Refinement methods provide a means of enhancing the granularity of our understanding, allowing for the exploration of finer details within the realm of lattice path structures.

Recursion serves as a powerful tool for describing and understanding the recursive nature inherent in many lattice path configurations, offering insights into their underlying principles and generating functions. Generating functionology, as elucidated by Wilf [Wil94], stands as a cornerstone in this area, providing a systematic framework for the study of generating functions and their applications in combinatorial analysis. Bijective methods offer an elegant approach to establishing one-to-one correspondences between different sets of lattice paths, shedding light on their structural properties and relationships.

While less frequently employed, the probabilistic method, as outlined by Gessel [Ges86], provides a valuable perspective by leveraging probabilistic reasoning to tackle combinatorial problems in the context of lattice paths. Additionally, Zeilberger's enumeration "ansatzes", introduced in [GBGL08], offers a rich source of strategies and techniques for enumerating lattice path configurations, blending insights from both enumerative and algebraic combinatorics. Through the collective utilization of these tools and methods, researchers in lattice path combinatorics are equipped with a versatile toolkit for exploring the myriad facets of this fascinating field.

The primary purpose of this monograph is to delve into the multifaceted properties of enumerative lattice path theory, an area of study experiencing rapid expansion at the forefront of combinatorics. Combinatorics itself is undergoing a substantial transformation, making this an opportune time to explore the intricacies of this domain. Our aim is to provide a comprehensive overview of several key topics within enumerative lattice path theory. Specifically, we will delve into combinatorial statistics, particularly those defined on lattice paths, exploring their unique characteristics and implications. Additionally, we will investigate special counting sequences and their q-analogs, shedding light on their significance within the context of lattice path theory. Furthermore, we will examine lattice paths in relation to their combinatorial nature, exploring their connections to broader issues such as general treatments, lattice path enumeration techniques, and the study of multiple paths. Through this exploration, we aim to offer insights into the rich tapestry of enumerative lattice path theory and its broader implications in combinatorics and beyond.

The "*Twelvefold Way*" to enumerate is a typical method in counting objects. Suppose we hope to investigate maps from a set A with cardinality n and a set B with cardinality m:

$$f : A \to B.$$

How many maps are there? To be simple, we have three considerations. (i) The set A contains n distinct elements, or all indistinguishable elements. (ii) The set B contains m distinct elements, or all indistinguishable elements. (iii) The maps are required to be all surjective, or all injective, or arbitrary. In total there are $2 \cdot 2 \cdot 3 = 12$ situations. The enumeration results of such maps are provided in Table 1.1. Here the point of "maps" is slightly different from usual in the sense that, if A contains n distinct elements $1, 2, \ldots, n$ but B contains all indistinguishable elements $y_1, y_2, \ldots, y_m$, and the map f is arbitrary, then we consider the following two maps f_1 and f_2 to be different, say,

$$f_1(i) = \begin{cases} y_1, & i = 1, 2; \\ y_2, & i \neq 1, 2; \end{cases}$$

and

$$f_2(i) = \begin{cases} y_1, & i = 1, 3; \\ y_2, & i \neq 1, 3. \end{cases}$$

This is because although y_1 and y_2 are indistinguishable in B, they are different images after all. The maps f_1 sends $1, 2 \in A$ to the same basket (image), but f_2 sends $1, 2 \in A$ to different baskets (images). Nonetheless the map f_3 is equivalent to f_1 since y_1 and y_2 are indistinguishable:

$$f_2(i) = \begin{cases} y_2, & i = 1, 2; \\ y_1, & i \neq 1, 2. \end{cases}$$

The full proofs are left to readers, or readers may check Stanley's classical book [Sta97, p31-40] for a wonderful discussion.

TABLE 1.1   The *Twelvefold Counting Formula*.

Elements of A	Elements of B	The Map f	Number of such Maps
all distinct	all distinct	arbitrary	m^n
all distinct	all distinct	injective	$(m)_n$
all distinct	all distinct	surjective	$m! S(n, m)$
all distinct	all indistinguishable	arbitrary	$\sum_{i=1}^{m} S(n, i)$
all distinct	all indistinguishable	injective	$\delta(n \leq m)$
all distinct	all indistinguishable	surjective	$S(n, m)$
all indistinguishable	all distinct	arbitrary	$\binom{n+m-1}{n}$
all indistinguishable	all distinct	injective	$\binom{m}{n}$
all indistinguishable	all distinct	surjective	$\binom{n-1}{m-1}$
all indistinguishable	all indistinguishable	arbitrary	$\sum_{i=1}^{m} p_i(n)$
all indistinguishable	all indistinguishable	injective	$\delta(n \leq m)$
all identical	all indistinguishable	surjective	$p_m(n)$

In Table 1.1, and throughout this monograph, $(m)_n := m(m-1)$ (see page 33), $S(n, k)$ represents *Stirling number of the second kind* (see page 33), $p_m(n)$ is the number of partitions of n into exactly m non-increasing parts (see page 42), and finally $\delta(P)$ is the Kronecker delta function defined by: $\delta(P) = 1$ if the statement P is true and $\delta(P) = 0$ if P is false.

To conclude this section, we address a little bit the idea of "combinatorial spirit". Combinatorics, as a mathematical discipline, has undergone continual evolution throughout history. In [Kan91], the author remarks that "[MacMahons] expertise lay in combinatorics, a sort of glorified dice-throwing, and in it he had made contributions original enough to be named a Fellow of the Royal Society". (see also [Cam]). Thus, the perception of combinatorialists and their contributions has not always been uniformly laudatory. However, "When Gian-Carlo Rota and various co-workers wrote an influential series of papers with the title ®On the foundations of combinatorial theory in the 1960s and 1970s, one reviewer compared combinatorialists to nomads on the steppes who had not managed to construct the cities in which other mathematicians dwell, and expressed the hope that these papers would at least found a thriving settlement. [WW13][p. 359]". As highlighted by Alon [Alo02], combinatorial theory has transcended its early reliance on ingenuity and detailed reasoning, evolving into a mature and multifaceted discipline, 'While many of the basic combinatorial results were obtained mainly by ingenuity and detailed reasoning, without relying on many deep, well developed tools, the modern theory has already grown out of this early stage [Alo02]".

Gowers et al suggests that enumeration is a foundational aspect of combinatorics, encapsulating its essence in the act of counting: "There are various ways in which one can try to define combinatorics. None is satisfactory on its own, but together they give some idea of what the subject is like. A first definition is that combinatorics is about counting things. [GBGL08][Part I, 2.7]". This sentiment is echoed by Benjamin and Quinn [BQ03], who demonstrate the power of simple counting arguments in understanding diverse number patterns thus broadening our perception of what constitutes "numbers that count".

In the author's perspective, proofs imbued with the combinatorial spirit often utilize constructions, bijections, and enumeration techniques. While other well-developed mathematical tools may be employed and play notable roles, it is the essence of enumeration and the systematic exploration of discrete structures that truly embodies the spirit of combinatorial mathematics. These proofs not only illuminate the elegance and ingenuity inherent in combinatorial reasoning but also contribute to the ongoing narrative of combinatorial theory's evolution and value within the broader mathematical landscape. One elementary example is the

Vandermonde's identity (or Vandermonde's convolution [Com74, page 44]): For all $n, m \geq 0$,

$$\binom{m+n}{k} = \sum_{i=0}^{k} \binom{m}{i}\binom{n}{k-i}.$$

Instead of proving the identity algebraically or using induction, a proof with combinatorial spirt is via the following interpretation. Suppose you have m unlabelled red balls and n unlabelled blue balls all of the same size inside of a bag. Now take randomly k balls out of the $m + n$ balls from the bag. The left side of Vandermonde's identity, $\binom{m+n}{k}$, is the total number of ways to take these k balls. While the right side, counts the number of ways to take k balls with precisely i red ones chosen from the m red balls (and forcibly $k - i$ blue ones taken from the n blues balls) for each i and adds these subcases up.

Remark 1.1.1. *In the proof above we used "double counting", namely counting the same set from two perspectives, and equalizing the formulas obtained respectively.*

In combinatorics, many of the most interesting proofs are *bijective*, that is, they provide explicit constructions of each side of a desired identity to be proved, and give a bijection between the two constructed sets. In [Sta], a list of almost 250 problems on bijective proofs, with around 27 open problems (as of August 2009).

1.2 RECURSIVE RELATIONS AND GENERATING FUNCTIONS

Recursive relations play a key role in the enumeration of many combinatorial objects. By a *combinatorial sequence*, we mean a sequence of integers $a_1, a_2, \ldots, a_n, \ldots$, such that a_n is the enumeration result of certain combinatorial objects with parameter n. For convenience, we often start with a_0 which is often 0 due to inherit consistence, In fact, this will be clear when we see how the term F_0 is decided in Example 1.2.2 below. Many combinatorial sequences satisfy certain recursive relations, that is, every term in previous terms of the sequence with a fixed pattern. Next we provide the definition of *linear recurrences*. More general patterns of recurrences may be found in [Bru04] and other classical texts in combinatorics.

Definition 1.2.1. *A sequence $\{a_n\}_{n\geq 0}$ satisfies a* linear recurrence of order k, *if there exist fixed constants $c_1, \ldots, c_k$ with $c_k \neq 0$, such that for all $n \geq k$, we have*

$$a_n = c_1 a_{n-1} + c_2 a_{n-2} + \cdots + c_k a_{n-k}. \tag{1.1}$$

Example 1.2.2. *The celebrated Fibonacci sequence $\{F_n\}_{n\geq 0}$, named after the Italian mathematician Leonardo Bonacci (also known as Fibonacci), cont for the number of pairs of rabbits in an ideal setting. Suppose at the beginning of the very first month there is a young pair of rabbits and at the of the first month, they mate, but there is still only one pair. At the end of the second month they produce a new pair, so there are two pairs in the field. At the end of the third month, the original pair produce a second pair, but the second pair only mate to gestate for a month, so there are three pairs in all... Furthermore suppose the rabbits never die. Let F_n be the number of pairs of rabbits at the beginning of the nth month, so that $F_1 = 1, F_2 = 1, F_3 = 2, F_4 = 3, \ldots$ It does not take much effort to observe that $\{F_n\}_{n\geq 0}$ satisfies a* linear recurrence of order 2*:*

$$F_n = F_{n-1} + F_{n-2}, \forall n \geq 2. \tag{1.2}$$

In order for (1.2) to work for $n = 2$, F_0 is taken to be 0.

In addition to the linear recurrence of order k as defined in Definition 1.2.1, there are nonlinear recurrences of a fixed order (see [Bru04] for a detailed discussion). Moreover, some combinatorial sequences have a recursive relation of order changing with the term, i.e., the nth term in the sequence is decided by $t(n)$ previous terms in some way, thus making it considerably more complicated. The generating function is a useful tool in dealing with this situation.

A generating function of a sequence is a formal power series that corresponds to this sequence. It provides a convenient way to record the sequence. There are various types of generating functions. Below, we only introduce two basic types briefly for the readers to get some flavors.

The *ordinary power series generating function (o.p.s.g.f. or often o.p.s.* for short) of sequence $\{a_n\}_{n\geq 0}$ is defined to be

$$f(x) = \sum_{n\geq 0} a_n x^n,$$

and the *exponential generating function (e.g.f.* for short) of sequence $\{a_n\}_{n\geq 0}$ is be

$$g(x) = \sum_{n\geq 0} \frac{a_n x^n}{n!}.$$

Example 1.2.3. *The ordinary power series generating function of the constant sequence $\{1\}_{n\geq 0}$ is*

$$f(x) = \sum_{n\geq 0} 1x^n = \frac{1}{1-x}.$$

The e.g.f. of the constant sequence $\{1\}_{n\geq 0}$ is

$$f(x) = \sum_{n\geq 0} \frac{a_n x^n}{n!} = e^x.$$

Example 1.2.4. *Let $F(x)$ be the ordinary power series generating function of the Fibonacci sequence. In (1.2), multiply with x^n and add up for $n \geq 2$, we get:*

$$F(x) - 0 - x = \sum_{n\geq 2} F_n x^n = \sum_{n\geq 2}(F_{n-1} + \sum_{n\geq 2} F_{n-2})x^n = x\sum_{i\geq 1} F_i x^i + x^2\sum_{j\geq 1} F_j x^j$$

$$= x(F(x) - 0) + x^2 F(x).$$

Solving the functional equation derived, we have $F(x) = \frac{x}{1-x-x^2}$. Within its radius of convergence, $F(x) = \frac{x}{1-x-x^2}$ may be expanded as follows,

$$\frac{x}{1-x-x^2} = \frac{-x}{x^2+x-1} = \frac{-x}{(x+\alpha)(x+\beta)}$$

$$= \frac{x}{\alpha-\beta}\left(\frac{1}{x+\alpha} - \frac{1}{x+\beta}\right)$$

$$= \frac{x}{\alpha-\beta}\left(\frac{1/\alpha}{1+x/\alpha} - \frac{1/\beta}{1+x/\beta}\right)$$

$$= \frac{x}{\alpha-\beta}\left(1/\alpha\sum_{i\geq 0}\frac{(-1)^i x^i}{\alpha^i} - 1/\beta\sum_{i\geq 0}\frac{(-1)^i x^i}{\beta^i}\right)$$

$$= \frac{x}{\sqrt{5}}\left(-\beta\sum_{i\geq 0}\beta^i x^i + \alpha\sum_{i\geq 0}\alpha^i x^i\right)$$

$$= \frac{x}{\sqrt{5}}\left(\sum_{i\geq 0}\alpha^{i+1} x^i - \sum_{i\geq 0}\beta^{i+1} x^i\right),$$

where $\alpha = \frac{1+\sqrt{5}}{2}$ is the golden ratio and $\beta = \frac{1-\sqrt{5}}{2} = \alpha - \sqrt{5} = \frac{-1}{\alpha}$ is its conjugate.

In general, by the symbol $[x^n]\{f(x)\}$ (or sometimes $[x^n]f(x)$ for convenience) we mean the coefficient of x^n in the formal power series $f(x)$ [Wil94]. Thus,

$$[x^n]\{F(x)\} = \frac{1}{\sqrt{5}}[x^{n-1}]\{(\sum_{i\geq 0}\alpha^{i+1}x^i - \sum_{i\geq 0}\beta^{i+1}x^i)\}$$

$$= \frac{1}{\sqrt{5}}(\alpha^n - \beta^n)$$

$$= \frac{1}{\sqrt{5}}((\frac{1+\sqrt{5}}{2})^n - (\frac{1-\sqrt{5}}{2})^n).$$

In Example 1.2.4, alternatively, the exponential generating function may be used. In that case formal differentiation will be needed and we will solve from a differential equation. It is often one's decision to choose whether the o.p.s. or the e.g.f. should be used, but in many cases both methods work after all.

To conclude this section we remark that in most cases we do have to worry about where the formal power series converge. As long as it converges, i.e. when the variable is within its convergence radius, we can manipulate the series and equalize coefficients of the appropriate expanded terms with the items of the combinatorial sequence.

For a wonderful resource of generating functions, [Wil94] is strongly recommended.

1.3 ORGANIZATION

This monograph is organized as follows.

In Chapter 2, after addressing the language of q-calculus, we introduce the theory of combinatorial statistics which is indispensable in the investigation of special counting sequences and lattice paths in future chapters. Mahonian and Eulerian statistics are of principal interests in our investigations, and the local or global property of combinatorial statistics is discussed.

Chapter 3 is devoted to discussions of special counting sequences including the Catalan numbers, Schröder numbers, their refinements and more general versions, Stirling numbers of the first and second kinds, Harmonic numbers, the integer partition numbers, as well as an introduction to the theory of Young tableaux with connections to combinatorial sequences.

In Chapter 4 we focus on lattice paths. Among others, we are mainly concerned with Dyck paths, Permutation paths, and Schröder paths. Lattice path enumerator is a helpful tool that counts the sum of $q^{\mathrm{area}(\Pi)}$ ranging over a set of lattice paths Π with a prescribed right boundary. We also treat with more general situations in that a lattice path may have arbitrary steps set and varied shapes of residence domain. Finally we briefly introduce noncrossing and nonintersecting multiple paths.

Some of the concepts and topics are interwound by nature. Aware of this fact, we try the best to weave together interconnected concepts and topics with care and precision. Our narrative unfolds in a logical sequence, designed to facilitate comprehension and foster a holistic understanding of the subject matter. We adopt two distinct approaches in our treatment: in many instances, we provide comprehensive details to elucidate the art and spirit of the theory and subjects under discussion. At other times, particularly when discussing seminal results such as Ramanujan's legendary congruences or recent cutting-edge advancements, we present the findings without proofs, offering references for readers to delve deeper into these fascinating areas of research. This dual approach ensures that readers can both appreciate the intricacies of the subject matter and explore avenues for further study and discovery.

CHAPTER 2

Combinatorial Statistics

Combinatorial statistics is an indispensable tool in the study of lattice path combinatorics and special counting sequences. We start with the q-calculus, the essential language of combinatorial statistics.

2.1 THE Q CALCULUS

The choice number or binomial coefficient $\binom{r}{k}$ may be extended to allow r to be any real number r and allow k to be any integer. Let $r \in \mathbb{R}$, $k \in \mathbb{Z}$, and we define the *generalized binomial coefficient* as follows.

$$\binom{r}{k} := \begin{cases} \frac{r(r-1)\cdots(r-k+1)}{k!}, & \text{if } k \geq 1, \\ 1, & \text{if } k = 0, \\ 0, & \text{if } k < 0. \end{cases}$$

It is easy to check that the above definition reduces to the ordinary choice number when n and k are both nonnegative integers.

We shall skip the proof of the following generalized binomial theorem credited to Sir Isaac Newton. The readers may refer to [Coo49] for a historical account and an analytical proof.

Theorem 2.1.1. *For any $-1 < x < 1$, $r \in \mathbb{R}$, we have*

$$(1+x)^r = \sum_{i=0}^{\infty} \binom{r}{i} x^i.$$

DOI: 10.1201/9781003509912-2

In combinatorics, a q-analogue of a counting function is typically a polynomial in q that simplifies to the original function when $q = 1$, and also retains some or all of the algebraic properties of the original function, like recursions.

Throughout this monograph we use the standard notation

$$[n] := (1 - q^n)/(1 - q), [n]! := [1][2] \cdots [n], \begin{bmatrix} n \\ k \end{bmatrix} := \frac{[n]!}{[k]![n - k]!}$$

for the q-analogue of the integer n, the q-factorial, and the q-binomial coefficient and $(a)_n := (1 - a)(1 - qa) \cdots (1 - q^{n-1}a)$ for the q-rising factorial. Sometimes it is necessary to write the base q explicitly as in $[n]_q$, $[n]!_q$, $\begin{bmatrix} n \\ k \end{bmatrix}_q$ and $(a; q)_n$, but we often omit q if it is clear from the context. Occasionally, when $i + j + k = n$, we also use $\begin{bmatrix} n \\ i,j,k \end{bmatrix} := \frac{[n]!}{[i]![j]![k]!}$ to represent the q-trinomial coefficient.

Sometimes the following "q-binomial theorem" is useful to prove q-identities.

Theorem 2.1.2. *The "q-binomial theorem". (See [And98, page 36].) For $n \in \mathbb{N}$,*

$$\sum_{k=0}^{n} q^{\binom{k}{2}} \begin{bmatrix} n \\ k \end{bmatrix} z^k = (-z; q)_n,$$

and

$$\sum_{k=0}^{\infty} \begin{bmatrix} n + k - 1 \\ k \end{bmatrix} z^k = \frac{1}{(z; q)_n}.$$

Another useful tool is the q-Vandermonde convolution, which may be obtained as a corollary of the q-binomial theorem.

Lemma 2.1.1. *The "q-Vandermonde convolution". (See [Hag08])*

$$\sum_{j=0}^{h} q^{(n-j)(h-j)} \begin{bmatrix} n \\ j \end{bmatrix} \begin{bmatrix} m \\ h - j \end{bmatrix} = \begin{bmatrix} m + n \\ h \end{bmatrix}.$$

The following q-identities are frequently utilized too, and are routine to verify.

Lemma 2.1.2. *(See [HS19].)*

(i) For nonnegative integers n, k, we have:

$$\begin{bmatrix} n \\ k - 1 \end{bmatrix} + q^k \begin{bmatrix} n \\ k \end{bmatrix} = \begin{bmatrix} n + 1 \\ k \end{bmatrix}. \tag{2.1}$$

(ii) For nonnegative integers $m \geq n$ and k, we have:

$$\begin{bmatrix} m \\ k \end{bmatrix} + q^{k+1} \begin{bmatrix} m-1 \\ k \end{bmatrix} + q^{2k+2} \begin{bmatrix} m-2 \\ k \end{bmatrix} + \cdots + q^{(m-n)(k+1)} \begin{bmatrix} n \\ k \end{bmatrix}$$
$$= \begin{bmatrix} m+1 \\ k+1 \end{bmatrix} - q^{(m-n+1)(k+1)} \begin{bmatrix} n \\ k+1 \end{bmatrix}.$$

(2.2)

2.2 MAHONIAN AND EULERIAN STATISTICS

A combinatorial statistic stat on a set S is a map $S \to \mathbb{N}$ with combinatorial significance in that the image $\mathrm{stat}(a)$ $(a \in S)$ computes the value of a with respect to a certain interesting property. The set S may be the symmetric group $\mathfrak{S}_n$, a set of words, or a set of lattice paths, as described in Chapter 4.

Three important statistics on words or permutations are inv, maj and des.

2.2.1 Mahonian Statistics

In general, for any integer word or multiset permutation $w = w_1 w_2 \cdots w_n$, the *inversion* and *major* statistics, denoted by inv and maj, respectively, are defined as

$$\mathrm{inv}(w) := \sum_{\substack{i<j \\ w_i > w_j}},$$

$$\mathrm{maj}(w) := \sum_{\substack{i \\ w_i > w_{i+1}}} i.$$

Related and also for later use, we define the *descent set* of a word w to be

$$Des(w) := \{i : w_i > w_{i+1}\},$$

and the *descent* statistic of w is its number of descents

$$\mathrm{des}(w) := |Des(w)|.$$

Example 2.2.1. *The* inv, maj, des *statistics of the permutation* $\sigma = \sigma_1\sigma_2\cdots\sigma_8 :=$ 53816274 $\in \mathfrak{S}_n$ *are, respectively,*

$$\mathrm{inv}(53816247) = 4 + 2 + 5 + 3 = 14$$

$$\mathrm{maj}(53816247) = 1 + 3 + 5 = 9$$

$$\mathrm{des}(53816247) = 1 + 1 + 1 = 3,$$

because that $5 = \sigma_1$ *generates 4 inversion pairs, with* $3, 1, 2, 4$ *appearing after 5, and* $3 = \sigma_2$ *generates 2 inversion pairs, etc, and that*

$$Des(\sigma) = \{1, 3, 5\}.$$

The following table shows the common statistics of all the permutations in $\mathfrak{S}_4$, *as well as their* descent *and* ascent *sets (for* asc *and* ascent set *see page 18).*

From Table 2.1, we may discover that

$$\sum_{\sigma \in \mathfrak{S}_4} q^{\mathrm{inv}(\sigma)} = \sum_{\sigma \in \mathfrak{S}_n} q^{\mathrm{maj}(\sigma)} = q^6 + 3q^5 + 5q^4 + 6q^3 + 5q^2 + 3q + 1. \qquad (2.3)$$

TABLE 2.1 Several combinatorial statistics and descent/ascent sets of $\sigma \in \mathfrak{S}_4$.

stat	1234	1243	1324	1342	1423	1432	2134	2143	2314	2341	2413	2431
inv	0	1	1	2	2	3	1	2	2	3	3	4
maj	0	3	2	3	2	5	1	4	2	3	2	5
des	0	1	1	1	1	2	1	2	1	1	1	2
asc	0	2	2	2	2	1	2	1	2	2	2	1
Des	empty	{3}	{2}	{3}	{2}	{2,3}	{1}	{1,3}	{2}	{3}	{2}	{2,3}
Asc	{1,2,3}	{1,2}	{1,3}	{1,2}	{1,3}	{1}	{2,3}	{2}	{1,3}	{1,2}	{1,3}	{1}
stat	3124	3142	3214	3241	3412	3421	4123	4132	4213	4231	4312	4321
inv	2	3	3	4	4	5	3	4	4	5	5	6
maj	1	4	3	4	2	5	1	4	3	4	3	6
des	1	2	2	2	1	2	1	2	2	2	2	3
asc	2	1	1	1	2	1	2	1	1	1	1	0
Des	{1}	{1,3}	{1,2}	{1,3}	{2}	{2,3}	{1}	{1,3}	{1,2}	{1,3}	{1,2}	{1,2,3}
Asc	{2,3}	{2}	{3}	{2}	{1,3}	{1}	{2,3}	{2}	{3}	{2}	{3}	empty

In fact, (2.3) is true for general n. The following two results due to MacMahon [Mac04] is now classical. We first look at inv and maj over the symmetric group and then investigate words.

Theorem 2.2.1. *[Mac04]*

$$\sum_{\sigma \in \mathfrak{S}_n} q^{\text{inv}(\sigma)} = [n]! = \sum_{\upsilon \in \mathfrak{S}_n} q^{\text{maj}(\sigma)}.$$

Proof. • $\sum_{\sigma \in \mathfrak{S}_n} q^{\text{inv}(\sigma)} = [n]$: Let $r(\sigma)$ be the number of letters in σ that appear after the letter n, and denote by σ^- the word (which is actually a permutation in $\mathfrak{S}_{n-1}$) obtained from σ by removing the letter n, so that $\text{inv}(\sigma) = r(\sigma) + \text{inv}(\sigma^-)$. It follows that,

$$\sum_{\sigma \in \mathfrak{S}_n} q^{\text{inv}(\sigma)} = \sum_{k=0}^{n-1} \sum_{\substack{\sigma \in \mathfrak{S}_n \\ r(\sigma) = k}} q^{\text{inv}(\sigma)} = \sum_{k=0}^{n-1} \sum_{\sigma^- \in \mathfrak{S}_{n-1}} q^{k + \text{inv}(\sigma^-)}$$

$$= \sum_{k=0}^{n-1} q^k \sum_{\sigma^- \in \mathfrak{S}_{n-1}} q^{\text{inv}(\sigma^-)} = [n] \sum_{\sigma \in \mathfrak{S}_{n-1}} q^{\text{inv}(\sigma)},$$

and inductively the conclusion is true.

• $\sum_{\sigma \in \mathfrak{S}_n} q^{\text{maj}(\sigma)} = [n]$: Observe that when the letter n is inserted in turn into each of the n possible spots of an arbitrarily fixed permutation $\tau \in \mathfrak{S}_{n-1}$, the increases of maj caused by the insertion exactly forms a traverse of $0, 1, \ldots, n-1$. So it is true by induction.

□

Next is the case of words.

Theorem 2.2.2. *For any integer s and fixed vector $\alpha \in \mathbb{N}^s$, if M_α denotes the set of all permutations of the multiset $\{1^{\alpha_1}, \ldots, s^{\alpha_s}\}$, then*

$$\sum_{w \in M_\alpha} q^{\text{inv}(w)} = \begin{bmatrix} n \\ \alpha_1, \ldots, \alpha_s \end{bmatrix} = \sum_{w \in M_\alpha} q^{\text{maj}(w)},$$

where $n = \sum_{i=1}^s \alpha_i$.

Proof. • $\displaystyle\sum_{w \in M_\alpha} q^{\mathrm{inv}(w)} = \begin{bmatrix} n \\ \alpha_1, \ldots, \alpha_s \end{bmatrix}$: Note that there is an obvious bijection between $M_\alpha \times \mathfrak{S}_{\alpha_1} \times \cdot \times \mathfrak{S}_{\alpha_s}$ and $\mathfrak{S}_n$, which preserves the value of inv statistic, i.e., $\mathrm{inv}(w) + \sum_{i=1}^{k} \mathrm{inv}(\sigma^{(i)}) = \mathrm{inv}(\sigma)$ ($w \in M_\alpha$, $\sigma^{(i)} \in \mathfrak{S}_{\alpha_i}$, $\sigma \in \mathfrak{S}_n$).

• $\displaystyle\sum_{w \in M_\alpha} q^{\mathrm{maj}(w)} = \begin{bmatrix} n \\ \alpha_1, \ldots, \alpha_s \end{bmatrix} = \begin{bmatrix} \alpha_1 \\ \alpha_1 \end{bmatrix} \begin{bmatrix} \alpha_1 + \alpha_2 \\ \alpha_2 \end{bmatrix} \cdots \begin{bmatrix} n \\ \alpha_s \end{bmatrix}$: Recursively, it suffices to show that $\displaystyle\sum_{w \in M_\alpha} q^{\mathrm{maj}(w)} = \sum_{u \in M_{(\alpha_1, \ldots, \alpha_{s-1})}} q^{\mathrm{maj}(u)} \begin{bmatrix} n \\ \alpha_s \end{bmatrix}$. In fact for any arbitrarily fixed $u \in M_{(\alpha_1, \ldots, \alpha_{s-1})}$, we prove the following two statements by induction.

(i) $\displaystyle\sum_{w \in u^+} q^{\mathrm{maj}(w) - \mathrm{maj}(u)} = \begin{bmatrix} n \\ \alpha_s \end{bmatrix}$, where w is taken over the $\binom{n}{\alpha_s}$ different words in M_α by adding α_s $s's$ to u;

and (ii) $\displaystyle\sum_{\substack{w \in u^+ \\ w_n \neq s}} q^{\mathrm{maj}(w) - \mathrm{maj}(u)} = \begin{bmatrix} n-1 \\ \alpha_s \end{bmatrix} q^{\alpha_s}$.

Observe that once (i) is established, the conclusion follows. Meanwhile by using (2.1), inductively, (i) is a corollary of (ii),

$$\sum_{w \in u^+} q^{\mathrm{maj}(w) - \mathrm{maj}(u)} = \sum_{\substack{w \in u^+; \\ w_n = s}} q^{\mathrm{maj}(w) - \mathrm{maj}(u)} + \sum_{\substack{w \in u^+; \\ w_n \neq s}} q^{\mathrm{maj}(w) - \mathrm{maj}(u)}$$

$$= \begin{bmatrix} n-1 \\ \alpha_s - 1 \end{bmatrix} + \begin{bmatrix} n-1 \\ \alpha_s \end{bmatrix} q^{\alpha_s} = \begin{bmatrix} n \\ \alpha_s \end{bmatrix}.$$

Nonetheless, the proof of (ii) is also based on (i) by an inductive logic.

Now let $l = n - \alpha_s$. In the case of (ii), as $w_n \neq s$, w_n is actually u_l pushed to the end with the α_s $s's$ added to all but the last spot in w. Let w^- (resp. u^-) represent the word obtained by deleting the last letter from w (resp. u). While w is taken over all the $\binom{n-1}{\alpha_s}$ words by adding α_s $s's$ to u with the requirement $w_n \neq s$, w^- is taken over all the $\binom{n-1}{\alpha_s}$ words by adding α_s $s's$ to u^- with no requirements.

Consider the following subcase in that $u_{l-1} > u_l = w_n$. Note that $(\mathrm{maj}(w) - \mathrm{maj}(u)) - (\mathrm{maj}(w^-) - \mathrm{maj}(u^-)) = (\mathrm{maj}(w) - \mathrm{maj}(w^-)) - (\mathrm{maj}(u) - \mathrm{maj}(u^-)) = (n-1) - (l-1) = \alpha_s$ whenever $w_{n-1} = u_{l-1}$ or $w_{n-1} = s$

(there are no other possibilities). Using the conclusion of (i) inductively, we get

$$\sum_{\substack{w \in u^+ \\ w_n \neq s}} q^{\text{maj}(w)-\text{maj}(u)} = \sum_{w^-} q^{\text{maj}(w^-)-\text{maj}(u^-)+\alpha_s} = \begin{bmatrix} n-1 \\ \alpha_s \end{bmatrix} q^{\alpha_s}.$$

The other subcase, i.e. $u_{l-1} \leq u_l = w_n$, is of similar spirit although more complicated. Let w^{--} denote the word obtained by deleting the last two letters from w. Then we have,

$$\sum_{\substack{w \in u^+ \\ w_n \neq s}} q^{\text{maj}(w)-\text{maj}(u)}$$

$$= \sum_{\substack{w \in u^+ \\ w_n = u_l \neq s; w_{n-1} = u_{l-1} \neq s}} q^{\text{maj}(w)-\text{maj}(u)} + \sum_{\substack{w \in u^+ \\ w_n \neq s; w_{n-1} = s}} q^{\text{maj}(w)-\text{maj}(u)}$$

$$= \sum_{\substack{w^-; w_{n-1}^- \neq s}} q^{\text{maj}(w^-)-\text{maj}(u^-)} + \sum_{w^{--}} q^{\text{maj}(w^{--})-\text{maj}(u^-)+(n-1)}$$

$$= \begin{bmatrix} n-2 \\ \alpha_s \end{bmatrix} q^{\alpha_s} + \begin{bmatrix} n-2 \\ \alpha_s - 1 \end{bmatrix} q^{n-1} = \begin{bmatrix} n-1 \\ \alpha_s \end{bmatrix} q^{\alpha_s}.$$

(The second to last step is based on inductive conclusions of (ii) and (i) in order. In the last step we have utilized (2.1) again.)

□

Accordingly, we say that inv and maj on M_α are *multiset Mahonian statistics*. Setting $s = n$ and $\alpha_1 = \cdots = \alpha_n = 1$ in the above theorem, M_α will be specialized to the symmetric group S_n, and Theorem 2.2.1 is rediscovered. Due to Theorem 2.2.1, we say that the two statistics inv and maj on S_n are both *Mahonian statistics*.

Two combinatorial statistics $\text{stat}_1, \text{stat}_2$ on the same set A are said to be *equivalently distributive* over A, "*equi-distribution*" for short, if they have the same distribution over A, namely

$$\sum_{a \in A} q^{\text{stat}_1(a)} = \sum_{a \in A} q^{\text{stat}_2(b)}.$$

Equi-distribution is denoted by "$stat_1 \equiv_A stat_2$". Sometimes "$\equiv_A$" is abbreviated to be "$\equiv$" if it is clear from the context. By Theorems 2.2.1 and 2.2.2,

$$\text{inv} \equiv_{\mathfrak{S}_n} \text{maj},$$

$$\text{inv} \equiv_{M_\alpha} \text{maj}.$$

2.2.2 Eulerian Statistics

Recall that the *descent* statistic is defined to be the cardinality of the descent set: $\text{des}(w) = |Des(w)|$. Symmetrically, the *ascent set* of a word w is defined to be

$$Asc(w) := \{i : w_i > w_{i+1}\},$$

and the *ascent* statistic is the cardinality of the ascent set: $\text{asc}(w) = |Asc(w)|$. On $\mathfrak{S}_n$, both des and asc are so-called Eulerian statistics. The asc and the ascent set of all permutations in $\mathfrak{S}_4$ are illustrated in Table 2.1.

The following sequence of summation polynomials taken over the symmetric group $\mathfrak{S}_n$

$$A_n(q) = \sum_{\sigma \in \mathfrak{S}_n} q^{1+\text{des}(\sigma)}$$

are called the *Eulerian polynomials*.

The coefficient of q^k in $A_n(q)$ is denoted $A_{n,k}$ and is called an *Eulerian number*. In other words, $A_{n,k} = |\mathcal{A}_{n,k}|$ where

$$\mathcal{A}_{n,k} = \{\sigma \in \mathfrak{S}_n : \text{des}(\sigma) = k - 1\},$$

so that $A_{n,k}$ counts the number of n-permutations with exactly $k - 1$ descents. The first few Eulerian polynomials are

$$A_1(q) = q$$
$$A_2(q) = q + q^2$$
$$A_3(q) = q + 4q^2 + q^3$$
$$A_4(q) = q + 11q^2 + 11q^3 + q^4$$

By convention, we set $A_0(q) = 1$.

Although there is no closed formula for $A_n(q) = \sum_{k=1}^{n} A_{n,k} q^k$, by discussing the position where letter n resides, it is convenient to find the following recurrence for the coefficients $A_{n,k}$: $A(n, k) = kA(n - 1, k) + (n - k + 1)A(n - 1, k - 1)$.

Also, the symmetry is clear: $A_{n,k} = A_{n,n-k+1}$ for $1 \leq k \leq n$.

Eulerian numbers and Eulerian polynomials were first introduced by Leonhard Euler in evaluating the power series with coefficients powers of fixed order. The following identity, of the nth power of a number as the sum of binomial coefficients of the nth order with coefficients the Eulerian numbers, is due to J. Worpitzky [Wor83]. (See, for example [Charalam, Chap. 14].)

Theorem 2.2.3. *(Worpitzky's Identity)*

$$x^n = \sum_{k=1}^n A_{n,k} \binom{x+n-k}{n}. \tag{2.4}$$

Using the symmetry relation $A_{n,k} = A_{n,n-k+1}$ for $1 \le k \le n$, one gets an equivalent form of (2.4),

$$x^n = \sum_{k=1}^n A_{n,k} \binom{x+k-1}{n}. \tag{2.5}$$

A q-analog of the identity (2.5) is given by Carlitz in [Car54]. Carlitz's q-Eulerian numbers, which we denote by $C_{n,k}(q)$, are defined by the expansion

$$[x]^n = \sum_{k=1}^n C_{n,k}(q) \begin{bmatrix} x+k-1 \\ n \end{bmatrix}. \qquad (n \ge 1) \tag{2.6}$$

In (2.6) $[x]$ is the q-analog of x as introduced in Section 2.1.

2.3 MISC: "LOCAL" VS. "GLOBAL"

Recall that two combinatorial statistics $\mathrm{stat}_1, \mathrm{stat}_2$ on the same set A are said to be *equivalently distributive* over A, if

$$\sum_{a \in A} q^{\mathrm{stat}_1(a)} = \sum_{a \in A} q^{\mathrm{stat}_2(b)}.$$

As a matter of fact, we could often prove the stronger fact of symmetry which readily implies the equi-distributive property:

$$\sum_{a \in A} q^{\mathrm{stat}_1(a)} t^{\mathrm{stat}_2(a)} = \sum_{a \in A} q^{\mathrm{stat}_2(a)} t^{\mathrm{stat}_1(a)}.$$

This is often achieved by constructing an involution ϕ on A with the requirement that $\mathrm{stat}_1(a) = \mathrm{stat}_2(\phi(a))$ for any $a \in A$. In fact such involutions are given by Foata and Schützenberger in their groundbreaking paper [FS78].

It is interesting that some statistics have the property that "local equals global" while others do not.

The *reduction* of any sequence of distinct integers $a = a_1 a_2 \cdots a_k$ is defined to be a permutation $\sigma = \sigma_1 \sigma_2 \cdots \sigma_k \in \mathfrak{S}_k$ such that $a_i < a_j$ if and only if $\sigma_i < \sigma_j$. This convention is both convenient and intuitive. Let $\underline{a}$ denote the reduction of a.

For example, for $a = 5394$, $\underline{a} = 3142$.

The ith *crown* of a, denoted by $\mathrm{crown}(a; i)$, is the reduction of $a_1 a_2 \cdots a_i$, i.e. $\underline{a_1 a_2 \cdots a_i}$. Likely, the ith *rear* of a, denoted by $\mathrm{rear}(a; i)$, is the reduction of $a_i a_{i+1} \cdots a_n$, i.e. $\underline{a_i a_{i+1} \cdots a_n}$. By definition, $\mathrm{crown}(a; i)$ is a permutation of length i, while $\mathrm{rear}(a; i)$ is a permutation of length $n - i + 1$. For $a = 5394$,

$$\mathrm{crown}(a; 2) = 21,$$

$$\mathrm{rear}(a; 2) = 132.$$

A permutation $\sigma \in \mathfrak{S}_n$ may be split into two shorter words pa and pb, both are distinct integer sequences over the alphabet $\{1, 2, \ldots, n\}$. More precisely, let i be arbitrarily chosen and fixed, and let $pa = \sigma_1 \sigma_2 \cdots \sigma_i$ and $pb = \sigma_{i+1} \cdots \sigma_n$.

For a certain permutation statistic stat, its *localization* at split location i is defined to be

$$\mathrm{stat}_{local}(\sigma, i) \triangleq \mathrm{stat}(\mathrm{crown}(\sigma; i)) + \mathrm{stat}(\mathrm{rear}(\sigma; i+1)) + \mathrm{stat}_{pair}(pa, pb),$$

$$(2.7)$$

where $\mathrm{stat}_{pair}(pa, pb)$ measures the mutual value of the split pairs prescribed by stat to be exemplified by examples later. We say that stat has the "*local-global property*" if its localized version stat_{local}, no matter where it splits (i.e. regardless of the split location i), is always equi-distributive with its global version (i.e. the original definition stat).

Explicitly, the "*local-global property*" says that for any $i = 1, \ldots, n$, it holds that

$$\mathrm{stat}_{local}(\sigma, i) \equiv_{\mathfrak{S}_n} \mathrm{stat}(\sigma).$$

For brevity in the following we just write $\underline{pa}$ for $\mathrm{crown}(\sigma; i)$ and $\underline{pb}$ for $\mathrm{rear}(\sigma; i+1)$.

In the above expression, stat_{pair}, being a reflection of the relationship between the two separated parts, mostly inherits the original statistic evaluated between the two parts. In fact, we may understand it this way:

$$\text{stat}(\sigma) = \text{stat}(pa) + \text{stat}(pb) + \text{stat}_{pair}(pa, pb). \tag{2.8}$$

Although the last terms of (2.7) and (2.8) are identical (stat_{pair}), as a critical component it should not be removed.

Example 2.3.1. *Let's start with* inv. *This is trivially true as* $\text{inv}(\sigma) = \text{inv}_{local}(\sigma, i)$ *is a universal fact: for any* $\sigma \in \mathfrak{S}_n$ *and any split location* i,

$$\text{inv}(pa) = \text{inv}(\underline{pa})$$
$$\text{inv}(pb) = \text{inv}(\underline{pb}),$$

as the relative order is preserved in the reduction.

Example 2.3.2. *The statistic* des *also has the local-global property.*

Next we explore sophisticated examples which are hopefully inspiring.

Example 2.3.3. *For* maj, *this is problematic. Although we could define* $\text{maj}_{pair}(pa, pb)$ *to be* i *when* $(pa)_i > (pb)_1$ *and* 0 *otherwise, the local stat is still lack of* $|pa|$ *times of* $\text{des}(pb)$ *compared to the global version of* maj. *This is difficult to get fixed naturally.*

However, let's consider the maj statistic applied to the inverse of a permutation. Originally, imaj is defined to be $\text{imaj}(\sigma) = \text{maj}(\sigma^{-1})$, i.e.,

$$\text{imaj}(\sigma) = \sum_{i:\ (\sigma^{-1})_i > (\sigma^{-1})_{i+1}} i.$$

But that is somehow inconvenient to use for our purpose. So we prefer to define it in the following way for technical reason. Instead let,

$$\text{iDes}(w) = \{i | i + 1 \text{ appears before } i \text{ in } w\}.$$

(Like the case of des we also define $\text{ides}(w) = |\text{iDes}(w)|$.)

Thus, $\text{imaj}(\sigma) = \sum_{i:\ i\in\text{iDes}(\sigma)} i$ appears to be a perfect analogue to the maj statistic.

By definition of iDes, to measures the mutual value of the split pairs is natural. We simply let

$$\text{iDes}_{pair}(pa, pb) = \{i | i + 1 \in pa \text{ and } i \text{ in } pb\}$$
$$\text{imaj}_{pair}(pa, pb) = \sum_{i:\ i\in\text{iDes}_{pair}(pa,pb)} .$$

Now consider

$$\text{imaj}_{local}(\sigma, i) = \text{imaj}(\underline{pa}) + \text{imaj}(\underline{pb}) + \text{imaj}_{pair}(pa, pb).$$

Is it equivalent, or at least equi-distributive to the global version of imaj?

Example 2.3.4. *Let* $\sigma = 539184267$. *Split at the 4th position, say. Thus* $pa = 5391$, $pb = 84267$, $\underline{pa} = 3241$ *and* $\underline{pb} = 52134$.

$$
\begin{aligned}
\text{imaj}_{local}(\sigma, 4) &= \text{imaj}(\underline{pa}) + \text{imaj}(\underline{pb}) + \text{imaj}(pa, pb) \\
&= \text{imaj}(3241) + \text{imaj}(52134) + \text{imaj}(5391, 84267) \\
&= (1 + 2) + (1 + 4) + (2 + 4 + 8) \\
&= 22; \\
\text{imaj}(\sigma) &= \text{imaj}(pa) + \text{imaj}(pb) + \text{imaj}(pa, pb) \\
&= \text{imaj}(5391) + \text{imaj}(84267) + \text{imaj}(5391, 84267) \\
&= (0) + (7) + (2 + 4 + 8) \\
&= 21.
\end{aligned}
$$

So imaj and imaj_{local} are not universally equal. Notice that iDes_{local}, i.e. the set of integers counted by imaj_{local}, with entries compressed during the process of reduction, tends to have more elements compared to iDes, i.e. the set of integers counted by imaj but the integers contained in iDes_{local} are generally smaller.

Theorem 2.3.1. *[Han]*

$$\text{imaj}_{local} \equiv_{\mathfrak{S}_n} \text{imaj}_{global}.$$

Proof. In the following we shall apply Foata's second transformation F_2 that exchanges inv and imaj [FS78]: $\mathrm{inv}(F_2(\sigma)) = \mathrm{maj}(\sigma)$ and $\mathrm{maj}(F_2(\sigma)) = \mathrm{inv}(\sigma)$ hold simultaneously, and will construct a sequence of compositions to construct a desired bijection.

For any $\sigma \in \mathfrak{S}_n$ and given split location, start with localized version

$$\mathrm{imaj}_{local}\sigma = \mathrm{imaj}(\underline{pa}) + \mathrm{imaj}(\underline{pb}) + \mathrm{imaj}_{pair}(pa, pb).$$

Applying Foata's second transformation, we get $\mathrm{imaj}_{local}\sigma = \mathrm{inv}(F_2(\underline{pa})) + \mathrm{inv}(F_2(\underline{pb})) + \mathrm{imaj}_{pair}(pa, pb)$. For simplicity let $S = \mathrm{imaj}_{local}\sigma$ and let $F_2(\underline{pa}) = fa$, $F_2(\underline{pb}) = fb$. In addition let h be the operation to rearrange integer sequences in non-decreasing order. Note that $S = \mathrm{inv}(fa) + \mathrm{inv}(fb) + \mathrm{imaj}_{pair}(pa, pb) = \mathrm{inv}(fa) + \mathrm{inv}(fb) + \mathrm{imaj}(h(pa), h(pb))$. Apply F_2 over $h(pa).h(pb) \in \mathfrak{S}_n$ (another permutation of length n), and name the image $ka.kb$, where $|ka| = |pa|$ and $|kb| = |pb|$. Thus $S = \mathrm{inv}(fa) + \mathrm{inv}(fb) + \mathrm{imaj}(h(pa).h(pb)) = \mathrm{inv}(fa) + \mathrm{inv}(fb) + \mathrm{inv}(ka.kb) = \mathrm{inv}(fa) + \mathrm{inv}(fb) + \mathrm{inv}(ka, kb)$. Above ka and kb are both increasing sequences. Reorder ka and kb according to the orders of fa and fb, keeping the alphabet of each, to obtain new words ra and rb, respectively. Therefore, $S = \mathrm{inv}(ra) + \mathrm{inv}(rb) + \mathrm{inv}(ra, rb) = \mathrm{inv}(ra.rb)$. Finally $S = \mathrm{imaj}(\tau)$ for $\tau = F_2^{-1}(ra.rb) \in \mathfrak{S}_n$. Every step is reversible so τ is uniquely decided, so that the local version and global (original) version are equi-distributive. □

CHAPTER 3

Special Counting Sequences

As commented by John J. Watkins in a review for MR3822822 [Slo18], "The world of mathematics contains many beautiful objects, and among the most beguiling are integer sequences, especially those sequences that hint at an order that remains always just beyond our grasp". In this chapter we introduce integer sequences that are of particular interests in contemporary combinatorics.

3.1 CATALAN NUMBERS, THE "MANIA"

There is a vast literature on Catalan families and related topics. See, for instance, [Cig87, FH85, FS78, FSV06, GG79, GH02, GH96, Ges92, Gou71, Hag03, Hai98, HP91, Knu73, Kra01, Kra89, Kre66, LSW11, LY90, Mac04, Moh79, Nar79, Rio68, Rio69, Son05b, Sta99, Sta15, Sul05]. In the present chapter we mainly use the language of words.

The nth *Catalan number* is

$$C_n = \frac{1}{n+1}\binom{2n}{n} = \frac{1}{2n+1}\binom{2n+1}{n}.$$

In [Sta15], a total of 214 kinds of objects counted by these numbers are presented, each of which is a different combinatorial interpretation of C_n. The phenomena of such a large amount of different objects being counted by C_n is sometimes called "Catalan disease" or "Catalania" (=Catalan mania) [Sta15, p. 56].

DOI: 10.1201/9781003509912-3

Definition 3.1.1. *Each **Catalan word** of order n is a word of length $2n$. It is a string of n 0's and n 1's, such that for every initial segment of the string, the number of 0's is at least as many as the number of 1's.*

Denote the set of Catalan words of order n by CW_n. Thus,

$$CW_3 = \{000111, 001011, 001101, 010011, 010101\}.$$

Theorem 3.1.1. *The nth Catalan number $C_n = 1, 2, 5, 14, 42, 132, 429, ...$ (OEIS: A000108) counts the number of Catalan words of order n. Moreover, C_n satisfies the recursive relation*

$$C_n = \sum_{k=1}^{n} C_{k-1} C_{n-k}. \tag{3.1}$$

Proof. Temporarily let $b_n = |CW_n|$. We derive a recurrence formula for b_n and show that b_n is indeed the Catalan number C_n. In fact, a Catalan word u of order k is called a **prime Catalan word** if for every initial segment of u other than the entire string itself, the number of 0's is always strictly more than the number of 1's. We may classify all Catalan words in CW_n according to the length of the first initial segment which is a prime Catalan word, and let that length be $2k$. Note that: (i) the number of prime Catalan words of order k is exactly the number of ordinary Catalan words of order $k - 1$, because removing the first 0 and last 1 from any prime Catalan word u (by definition u is required to start with 0 and end with 1), we get an ordinary Catalan word of lower order, and vice versa; (ii) Removing the initial segment from $w \in CW_n$ which is a prime Catalan word of order k would leave an ordinary Catalan word of order $n - k$. Hence, $b_n = \sum_{k=1}^{n} b_{k-1} b_{n-k}$.

Next we show that $b_n = \frac{1}{n+1} \binom{2n}{n}$. Among the $\binom{2n}{n}$ words with exactly n 0's and n 1's, we enumerate those non-Catalan words by a bijection. Let $v = v_1 \cdots v_{2n}$ be such a non-Catalan word. Suppose at position $2k+1$, for the first time the initial segment $v_1 \cdots v_{2k+1}$ has more 1's than 0's. Notice that $v_1 \cdots v_{2k}$ is a Catalan word of order k and that $v_{2k+1} = 1$. Now map v to $y = v_1 \cdots v_{2k+1}(1 - v_{2k+2}) \cdots (1 - v_{2n})$. Then y is a word with exactly $n - 1$ 0's and $n + 1$ 1's. Reversely, for every $\{0,1\}$ word y with exactly $n - 1$ 0's and $n + 1$ 1's, we locate the uniquely decided maximum initial string u such that for every initial substring of u the number of 0's is at least as many as the number of 1's. Keep u in y, keep the next 1 right after u, and reflect the part of y after that 1 which has initially one more 1's than 0's, then the string y is mapped back to a non-Catalan word of n 0's and n 1's. Because there are in total $\binom{2n}{n+1}$ words with exactly $n-1$ 0's and $n+1$ 1's, $b_n = \binom{2n}{n} - \binom{2n}{n+1} = \frac{1}{2n}\binom{2n}{n} = C_n$. □

Remark 3.1.2. *In the above proof we have basically followed the spirit of André's Reflection Method [And87].*

Letting $n = 1$ in (3.1), it makes sense to set $C_n = 1$.

Among the many interpretations, we are primarily interested in those associated to the lattice paths, which will be discussed in the next chapter.

Theorem 3.1.2. *The ordinary generating function of $\{C_n\}_{n=0}^\infty$, where $C_0 = 1$, is the following:*

$$\frac{1 - \sqrt{1 - 4x}}{2x}.$$

Proof. Let $C(x) = \sum_{n=0}^\infty C_n x^n$. For convenience, set $a_n = C_{n-1}$, $n \geq 1$, and $a_0 = 0$. So for any $n \geq 1$, $C_n = \sum_{k=1}^n C_{k-1} C_{n-k} = \sum_{k=0}^n a_k C_{n-k}$. This implies $C(x) - 1 = (\sum_{n \geq 0} a_n x^n) C(x) = xC(x)$. Solving the functional equation, we get

$$C(x) = \frac{1 - \sqrt{1 - 4x}}{2x}.$$ □

Carlitz and Riordan [CR64] defined the following natural q-analogue of C_n,

$$C_n(q) = \sum_{w \in CW_n} q^{\text{inv}(w)}, \tag{3.2}$$

and showed that

Theorem 3.1.3.

$$C_n(q) = q^{k-1} \sum_{k=1}^n C_{k-1}(q) C_{n-k}(q), n \geq 1.$$

Unfortunately the above Carlitz and Riordan's q-Catalan numbers have no closed form expressions.

Nevertheless, if we replace the power inv by maj in each monomial term of 3.2, the summation turns out to have an elegant formula. The following classical result of MacMahon [Mac60, p. 214] has a simple combinatorial proof in [FH85, (3.5)]. The spirit of Fürlinger and Hofbauer's proof is similar to what we have showed in the proof of Theorem 3.1.1.

Theorem 3.1.4.

$$\sum_{w \in CW_n} q^{\mathrm{maj}(w)} = \frac{1}{[n+1]} \begin{bmatrix} 2n \\ n \end{bmatrix}.$$

3.2 REFINEMENTS AND EXTENSIONS OF CATALAN NUMBERS

3.2.1 Narayana Numbers

In 1950's Narayana ([Nar55], see also [Nar59, Nar79]) and Runyon (see [Rio68]) independently introduced a nice refinement of Catalan numbers. Basically, it counts the number of Catalan words with fixed number of descents (or ascents),

$$N_{n,k} = \frac{1}{n} \binom{n}{k} \binom{n}{k+1}.$$

In 1985, Fürlinger and Hofbauer [FH85] investigated the two types of q-Catalan numbers mentioned in the previous section, and proved several useful results involving statistics on the lattice paths, including the following combinatorial interpretation of q-analogs of Narayana numbers (or Runyon numbers),

$$\sum_{w \in CW_n : \mathrm{des}(w) = k} q^{\mathrm{maj}(w)} = \frac{1}{[n]} \begin{bmatrix} n \\ k \end{bmatrix} \begin{bmatrix} n \\ k+1 \end{bmatrix} q^{k^2+k}. \tag{3.3}$$

([FH85, Equation 4.1]; see also [Mac04, p. 2429].) Note that (3.3) is a refinement of Theorem 3.1.4, in the sense that the domain of the summation CW_n is classified according to the number of descents, $(\mathrm{des}(w) = k$, or equivalently $\mathrm{asc}(w) = k+1)$.

3.2.2 Super Catalan Numbers

Catalan numbers are generalized in several directions.

In one direction, the Catalan numbers expressed by $C_n = \frac{1}{2n+1} \binom{2n+1}{n}$ generalize readily to the ballot numbers,

$$b(n, k) = \frac{k}{2n+k} \binom{2n+k}{n}. \tag{3.4}$$

The classical ballot problem solved first by Bertrand [Ber72] is the following. Suppose Alice and Bob are candidates for office and suppose that among the $2n + k$ $(n, k \geq 1)$ voters, $n+k$ of them vote for Alice while n vote for Bob. In how many ways can the ballots be counted so that at every stage Alice stays ahead of Bob? The solution is the ballot number $b(n, k)$ in (3.4). Clearly, $b(n + 1, n) = C_n$. In fact, if we interpret each acceptable outcome in the ballot problem with parameter $(n + 1, n)$ to be a sequence of $n + 1$ 0's and n 1's, then removing the initial 0 gives us exactly a sequence in CW_n, and vice versa.

In 1874, Catalan stated that the numbers

$$S(n, m) = \frac{(2m)!(2n)!}{m!n!(m + n)!}$$

are integers. (See [Ges92, footnote 1].) $S(n, m)$ are called the *super Catalan numbers* by Gessel, as $S(n, 1)/2 = C_n$, $S(n, 2)/2 = 4C_n - C_{n+1}$, and more generally $S(n, m)/2$ is the *super ballot number* $g(n, 1, m - 1)$ [Ges92]. In general, it remains an intriguing open problem to find a direct combinatorial interpretation of the super Catalan numbers [CW].

We adopt the definition of generalized binomial coefficient $\binom{r}{k}$ as in Section 2.1. For positive integers p, q with $q < p$, the Patalan number $b(n; p, q) = -p^{2n+1}\binom{n-\frac{q}{p}}{n+1}$ is defined in [Ric15]. Note that $b(n; 2, 1)$ is equivalent to the Catalan numbers $C_n = \frac{2n!}{n!(n+1)!}$. Meanwhile, the $(n, m; p, q)$-super Patalan number is

$$Q(n, m; p, q) = (-1)^m p^{2(n+m)} \binom{n - \frac{q}{p}}{m + n}.$$

It is proved that the super Patalan numbers are integers and that they generalize the super Catalan numbers $S(n, m)$ in the same way that the Patalan numbers generalize the classical Catalan numbers: $Q(n, m; 2, 1) = S(n, m)$ [Ric15].

3.2.3 Fuss-Catalan Numbers

The m-Catalan number C_n^m studied by Cigler counts the *Catalan words of order n and dimension m*,

$$CW_n^m = \{w \mid \begin{array}{l} w \text{ is a string of n 0'a and } mn \text{ 1's, such that at any initial segment of } w, \\ m \text{ times the number of 0's is at least as many as the number of 1's.} \end{array}\}.$$

He showed that

$$|CW_n^m| = \frac{1}{mn+1}\binom{mn+n}{n}.\text{([Cig87], see also [HP91] and [HPW99]))}$$

In fact, Cigler [Cig87] proved that the number of Catalan words of order n and dimension m with exactly k descents (or equivalently $k+1$ ascents) is counted by the generalized Narayana number,

$$N_{n,k}^m = \frac{1}{n}\binom{n}{k+1}\binom{mn}{k}. \tag{3.5}$$

If we go further to sum these words up with respect to the maj statistic, we wouldn't get a generalization of (3.3), however. While

$$\sum_{w\in CW_n^m} q^{\text{maj}(w)} = \sum_k \sum_{w\in CW_n^m:\text{des}(w)=k} q^{\text{maj}(w)},$$

and using the q-Vandermonde convolution (Lemma 2.1.1),

$$\sum_k \frac{1}{[n]}\begin{bmatrix}mn\\k\end{bmatrix}\begin{bmatrix}n\\k+1\end{bmatrix}q^{k^2+k} = \frac{1}{[n]}\begin{bmatrix}mn+n\\n-1\end{bmatrix} = \frac{1}{[mn+1]}\begin{bmatrix}mn+n\\n\end{bmatrix},$$

the conceivable equality $\sum_{w\in CW_n^m} q^{\text{maj}(w)} = \frac{1}{[mn+1]}\begin{bmatrix}mn+n\\n\end{bmatrix}$ does not hold in general. For instance, let $n=2$ and $m=2$, we get

$$\sum_{w\in CW_n^m} q^{\text{maj}(w)} = 1+q^2+q^3 \neq 1+q^2+q^4 == \frac{1}{[mn+1]}\begin{bmatrix}mn+n\\n\end{bmatrix}.$$

The m-Catalan numbers are also called Fuss-Catalan numbers [Ava08]. In [LSW11], an explicit formula is found for the density of the probability measures π_m having as moments the Fuss-Catalan numbers $\frac{1}{mn+1}\binom{mn+n}{n}$.

3.2.4 Schröder Numbers and Delannoy Numbers

Introduced by Ernst Schröder [Sch70], the nth *Schröder number* is

$$S_n = \sum_{d=0}^n \frac{1}{n-d+1}\binom{2n-d}{d,n-d,n-d}. \tag{3.6}$$

The nth Schröder number $S_n = 2, 6, 22, 90, 394, 1806, 8558, \ldots$ (OEIS: A006318) counts the number of Schröder words of order n introduced below.

Definition 3.2.1. *Each d-**Schröder word** of order n is a word of length $2n + d$, consisted of $n - d$ 0's, d 1's, and $n - d$ 2's, such that for every initial segment of the string, the number of 0's is at least the number of 2's. We denote the set of d-Schröder words of order n by $SW_{n,d}$.*

The set of **Schröder words of order n**, denoted by SW_n, is the union of d-Schröder words of order n for all appropriate d's.

For example,

$$SW_3 = \bigcup_{d=0}^{3} SW_{n,d}$$

$$= \{000222, 002022, 002202, 020022, 020202, 10022, 01022, 00122,$$

$$00212, 00221, 1102, 1012, 1021, 0112, 0120, 0211, 111\}.$$

By definition, $SW_{n,0} = CW_n$ and so $|SW_{n,0}|$ is exactly the nth Catalan number. In addition, observe that by usual combinatorial argument $|SW_{n,d}| = \binom{2n-d}{d} C_{n-d}$, and therefore, $|SW_{n,d}| = \frac{1}{n-d+1} \binom{2n-d}{d, n-d, n-d}$ follows. Hence the cardinality of SW_n is S_n.

Note that $S_0 = 1$. Momentarily let k be the first occasion such that in a Schröder word of order n, the number of 2's (which is k) for the fist time matches the number of 0's, with the possibility $k = 0$ in which case the word starts with letter 1. Thereupon the following recursive relation is easy to observe,

$$S_n = S_{n-1} + \sum_{k=1}^{n} S_{k-1} S_{n-k} = S_{n-1} + \sum_{k=1}^{n} S_{k-1} S_{n-k}. \tag{3.7}$$

Thus the ordinary generating function of $\{S_n\}_{n \geq 0}$ may be derived from (3.7):

$$S(x) = \frac{1 - x - \sqrt{1 - 6x + x^2}}{2x}.$$

Sometimes S_n is called the large Schröder number or big Schröder number, to distinguish from $s_n = \frac{S_n}{2}$ (OEIS: A001003), which is named the little Schröder number or small Schröder number. [Sta99, Ex.6.39, p. 239-240] provides 19 combinatorial interpretations of these numbers.

For every Schröder word, we may define the maj statistic as well. For example, $\mathrm{maj}(w) = 5 + 6 + 8 + 11 = 30$ for $w = 001221010221 \in SW_{8,4}$.

In Bonin, Shapiro's and Simion's classical work, they proved a graceful result about the maj statistic and the q-Schröder number.

Theorem 3.2.1. *[BSS93, Theorem 4.3]*

$$\sum_{w \in SW_{n,d}} q^{\mathrm{maj}(w)} = \frac{1}{[n-d+1]} \begin{bmatrix} 2n-d \\ n-d, n-d, d \end{bmatrix}.$$

Henri Delannoy introduced what we now call the Delannoy numbers or the Delannoy array [Del95]. The (m,n)-*Delannoy numbers* are $D(m,n) := \sum_{d \geq 0} D(m,n,d)$, where

$$D(m,n,d) = \binom{m+n-d}{m-d, n-d, d}.$$

They satisfy the recurrence $D(m,n) = D(m-1,n) + D(m,n-1) + D(m-1,n-1)$, and the generating function [Com74, p. 81]

$$\sum_{m,n \geq 0} D(m,n)x^m y^n = (1 - x - y - xy)^{-1}.$$

When $m = n$, $D(n,n) = \sum_{d \geq 0} D(n,n,d)$ are the central-Delannoy numbers. Note that

$$\frac{S(n,d)}{D(n,n,d)} = \frac{1}{n-d+1},$$

is a generalized formula of $C_n / \binom{2n}{n} = \frac{1}{n+1}$, where d is taken to be 0.

Among others, [Sul03] is a wonderful resource containing a catalog of 29 configurations counted by the Delannoy numbers.

3.3 STIRLING NUMBERS, HARMONIC NUMBERS, ETC.

The special counting sequences to be discussed below are interesting and important in combinatorics and other areas. Unless those introduced in earlier sections, though, the number here are less relevant to lattice paths.

3.3.1   Stirling Numbers and Bell Numbers

Definition 3.3.1. *For $n, k \in \mathbb{Z}^+$, $c(n, k)$ represents the number of permutations in the symmetric group $\mathfrak{S}_n$ with exactly k cycles. The* Stirling number of the first kind *is $s(n, k) = (-1)^{n-k} c(n, k)$. We also say that $c(n, k)$ is the* unsigned *Stirling number of the first kind.*

It makes sense to set $c(0, 0) = 1$ and $c(n, 0) = c(0, n) = 0$ for $n \geq 1$.

Lemma 3.3.2. *For $n, k \geq 1$, $c(n, k)$ satisfies the recurrence*

$$c(n, k) = (n - 1)c(n - 1, k) + c(n - 1, k - 1).$$

Proof. Take $\sigma \in \mathfrak{S}_n$ with exactly k cycles in its cyclic decomposition. If $\sigma_n = n$, then (n) is a cycle in σ. So the number of such $\sigma's$ is equivalent to the number of permutations in $\mathfrak{S}_{n-1}$ with exactly $k - 1$ cycles, which is counted by $c(n - 1, k - 1)$. If $\sigma_n \neq n$, then removing n from the cyclic decomposition of σ leaves a permutation in $\mathfrak{S}_{n-1}$ with k nonempty cycles; while adding n to any permutation in $\mathfrak{S}_{n-1}$ with k cycles, there will be $n - 1$ ways to create a permutation in $\mathfrak{S}_n$ without changing the number of cycles (decided by the element of $\{1, \ldots, n - 1\}$ that is preceding n in the same cycle of cyclic decomposition). So it is true. □

Theorem 3.3.1. *$\{c(n, k)\}_{n=1}^{\infty}$ satisfies the following generating function or functional equation*

$$\sum_{k=1}^{n} c(n, k)x^k = x(x + 1) \cdots (x + n - 1).$$

Proof. Recall that by the symbol $[x^n]\{f(x)\}$ we mean the coefficient of x^n in the series $f(x)$. Inductively,

$$
\begin{aligned}
&[x^k]\{x(x + 1) \cdots (x + n - 1)\} \\
&= [x^k]\{x(x + 1) \cdots (x + n - 2)x\} + (n - 1)[x^k]\{x(x + 1) \cdots (x + n - 2)\} \\
&= [x^{k-1}]\{x(x + 1) \cdots (x + n - 2)\} + (n - 1)[x^k]\{x(x + 1) \cdots (x + n - 2)\} \\
&= c(n - 1, k - 1) + (n - 1)c(n - 1, k) \\
&= c(n, k),
\end{aligned}
$$

where the second to last equality is based on induction hypothesis. □

Remark 3.3.3. *Sometimes a different convention is used. In place of* $s(n, k)$, $c(n, k) = (-1)^{n-k}s(n, k)$, *i.e., Stirling number of the first kind with no signs defined in Definition 3.3.1, are named the Stirling numbers of the first kind instead. Although the numbers* $s(n, k)$ *are sometimes negative, they are essentially more convenient, so we prefer the current definition. There are also other notation for the Stirling numbers such as* $s_1(n, k), s_2(n, k)$ *and* $\left[{n \atop k}\right], \left\{{n \atop k}\right\}$. *See [vLW92, Chap. 13] and [GKP89, Sec 6.1].*

The following is a dual form of Theorem 3.3.1.

Theorem 3.3.2. $\{s(n, k)\}_{n=1}^{\infty}$ *satisfies the functional equation*

$$\sum_{k=1}^{n} s(n, k)x^k = (x)_n.$$

Here $(x)_n = x(x - 1) \cdots (x - n + 1)$.

Proof. Simply replace x by $-x$ in Theorem 3.3.1, and multiply by $(-1)^n$ on both sides.

$$(-1)^n \sum_{k=1}^{n} c(n, k)(-x)^k = (-1)^n(-x)(-x + 1) \cdots (-x + n - 1),$$

gives exactly the desired identity. □

Definition 3.3.4. *For* $n, k \in \mathbb{N}$, *denoted by* $S(n, k)$, *the* Stirling number of the second kind *is the number of partitions of* $\{1, 2, \ldots, n\}$ *into* k *nonempty subsets.*

Set $S(n, 0) = S(0, n) = 0$ for $n \geq 1$, and $S(0, 0) = 1$.

Stirling numbers of the second kind satisfy a recursive relation similar to the one given in Lemma 3.3.2.

Lemma 3.3.5. *For* $n, k \geq 1$, $S(n, k)$ *is decided by*

$$S(n, k) = kS(n - 1, k) + S(n - 1, k - 1).$$

Proof. (*Sketch.*) Let the k-partitions of $\{1, 2, \ldots, n\}$ be $\mathfrak{P}(n, k)$. Map $\mathfrak{P}(n, k)$ bijectively to $k\mathfrak{P}(n - 1, k) \cup \mathfrak{P}(n - 1, k - 1)$ according to where the element n is partitioned to. ($k\mathfrak{P}(n - 1, k)$ means k copies of $\mathfrak{P}(n - 1, k)$ labelled $1, 2, \ldots, k$.) If

n is by itself, the image of the partition $\mathcal{P}$ in $\mathfrak{P}(n, k)$ is $\mathfrak{P}(n-1, k-1)$. Otherwise, the partition $\mathcal{P}$ is mapped to a copy of $\mathfrak{P}(n-1, k)$ decided by the specific one of the k subsets in $\mathcal{P}$ that contains n. □

Theorem 3.3.3. $\{S(n, k)\}_{n=1}^{\infty}$ *satisfies the functional equation*

$$\sum_{k=1}^{n} S(n, k)(x)_k = x^n.$$

Here $(x)_k = x(x-1)\cdots(x-k+1)$.

Proof. By induction and Lemma 3.3.5,

$$x^n = x^{n-1}x = \sum_{k=1}^{n-1} S(n-1, k)(x)_k x$$

$$= \sum_{k=1}^{n-1} S(n-1, k)(x)_k(x-k+k)$$

$$= \sum_{k=1}^{n-1} S(n-1, k)(x)_{k+1} + \sum_{k=1}^{n-1} kS(n-1, k)(x)_k$$

$$= \sum_{k=2}^{n} S(n-1, k-1)(x)_k + \sum_{k=1}^{n-1} kS(n-1, k)(x)_k$$

$$= \sum_{k=1}^{n} S(n-1, k-1)(x)_k + \sum_{k=1}^{n} kS(n-1, k)(x)_k$$

$$= \sum_{k=1}^{n} S(n, k)(x)_k.$$

□

Remark 3.3.6. *Theorem 3.3.3 may be interpreted as follows. Suppose x is a positive integer. In total there are x^n maps from $[n]$ to $[x]$. These maps may be surjections to $[x]$ or not, but each of them is a surjection to a uniquely decided k-subset of $[x]$. Meanwhile for every k-subset $Y \subseteq [x]$, the number of surjections from $[n]$ to Y is $k!S(n, k)$. Hence we have*

$$x^n = \sum_{k=1}^{n} \binom{x}{k} k!S(n, k) = \sum_{k=1}^{n} S(n, k)(x)_k.$$

Theorem 3.3.4. *The n by n matrices with two types of Stirling numbers as entries, respectively, are inverses to each other. That is, define $A = (a_{ij})_{n \times n} := (s(i,j))_{n \times n}$ and $B = (b_{ij})_{n \times n} := (S(i,j))_{n \times n}$. Then*

$$AB = BA = I.$$

Proof. Note that $\{x, x^2, \ldots, x^n\}$ and $\{(x)_1, (x)_2, \ldots, (x)_n\}$ are both vector bases of the vector space of all zero-constant polynomials of order no more than n over any field, $\{\sum_{i=1}^n \lambda_i x^i \mid \lambda_i \in \mathbb{C}\}$ (say, over $\mathbb{C}$). A and B are just transition matrices between the two vector bases. $\square$

Corollary 3.3.7.

$$\sum_{l \geq 0}^n s(i,l)S(l,j) = \delta(i,j),$$

$$\sum_{l \geq 0}^n S(i,l)s(l,j) = \delta(i,j).$$

(Throughout this monograph $\delta(x,y)$ is the Kronecker delta function defined by: $\delta(x,y) = 1$ for $x = y$ and $\delta(x,y) = 0$ for $x \neq y$.)

On another perspective, by using the Inclusion-Exclusion Principle, we have the following identity.

Lemma 3.3.8. *For any positive integers n, k, we have*

$$S(n,k) = \frac{1}{k!} \sum_{j=0}^k \binom{k}{j} j^n (-1)^{k-j}.$$

Proof. Note that $k!S(n,k)$ represent the number of surjective maps from $\{1, \ldots, n\}$ to $\{1, \ldots, k\}$. Let A_i be the collection of maps $\{1, \ldots, n\}$ to $\{1, \ldots, k\}$ such that $i \in \{1, \ldots, k\}$ is not mapped to. Hence,

$$k!S(n,k) = |\bigcap_{1 \leq i \leq k} \overline{A_i}| = k^n - \sum_i |A_i| + \sum_{1 \leq i < j \leq k} |A_i \cap A_j|$$

$$- \sum_{1 \leq i < j < t \leq k} |A_i \cap A_j \cap A_l| + \cdots + (-1)^k |A_1 \cap A_2 \cap \cdots \cap A_k|$$

$$= \sum_{r=0}^k \binom{k}{r} (k-r)^n (-1)^r$$

$$= \sum_{j=0}^k \binom{k}{j} j^n (-1)^{k-j}. \qquad \square$$

Now it is ready to discover the exponential generating function of $\{S(n,k)\}_{n\geq 0}$.

Theorem 3.3.5. *The exponential generating function of* $\{S(n,k)\}_{n=0}^{\infty}$ *is*

$$\sum_{n=0}^{\infty} \frac{S(n,k)x^n}{n!} = \frac{(e^x - 1)^k}{k!}.$$

Proof. Straightforwardly,

$$\sum_{n=0}^{\infty} \frac{S(n,k)x^n}{n!} = \sum_{n=0}^{\infty} \frac{1}{k!} \sum_{j=0}^{k} \binom{k}{j} j^n (-1)^{k-j} \frac{x^n}{n!}$$

$$= \sum_{j=0}^{k} (-1)^{k-j} \frac{1}{k!} \binom{k}{j} \sum_{n=0}^{\infty} j^n \frac{x^n}{n!}$$

$$= \sum_{j=0}^{k} (-1)^{k-j} \frac{1}{k!} \binom{k}{j} e^{jx}$$

$$= \frac{1}{k!} \sum_{j=0}^{k} \binom{k}{j} (e^x)^j (-1)^{k-j}$$

$$= \frac{1}{k!} (e^x - 1)^k.$$

$\square$

Next, by utilizing Theorem 3.3.5, we shall find the exponential generating function of Stirling numbers of the first kind. Ahead of that we need to introduce the following tool.

Lemma 3.3.9. *Let* $A(x)$ *and* $B(x)$ *be the exponential generating functions of sequences* $\{a_n\}_{n=0}^{\infty}$ *and* $\{b_n\}_{n=0}^{\infty}$*, respectively. The following statements are equivalent.*

(i) *For all* $n \geq 0$, $a_n = \sum_{i\geq 0} s(n,i)b_i$;

(ii) *For all* $n \geq 0$, $b_n = \sum_{i\geq 0} S(n,i)a_i$;

(iii) $B(x) = A(e^x - 1)$, *i.e.,* $A(x) = B(\ln(1+x))$.

Proof. We only demonstrate that (i) $\Rightarrow$ (ii) $\Rightarrow$ (iii), and leave to the reader to verify the rest.

(i) implies (ii): Assuming (i), by Corollary 3.3.7,

$$
\begin{aligned}
\sum_{j\geq 0} S(n,j)a_j &= \sum_{j\geq 0} S(n,j) \sum_{i\geq 0} s(j,i)b_i \\
&= \sum_{i\geq 0} b_i \sum_{j\geq 0} S(n,j)s(j,i) \\
&= \sum_{i\geq 0} b_i \delta(n,i) = b_n.
\end{aligned}
$$

(ii) implies (iii): Assuming (ii), by definition,

$$
\begin{aligned}
B(x) = \sum_{n\geq 0} b_n \frac{x^n}{n!} &= \sum_{n\geq 0}\sum_{i\geq 0} S(n,i)a_i \frac{x^n}{n!} \\
&= \sum_{i\geq 0} a_i \sum_{n\geq 0} S(n,i)\frac{x^n}{n!} \\
&= \sum_{i\geq 0} a_i \frac{(e^x-1)^i}{i!} = A(e^x-1).
\end{aligned}
$$

$\square$

Theorem 3.3.6. *The exponential generating function of* $\{s(n,k)\}_{n=0}^{\infty}$ *is*

$$
\frac{(\ln(1+x))^k}{k!}.
$$

Proof. For $n \in \mathbb{N}$, let $a_n = s(n,k)$ and $b_n = \delta(n,k)$. Thus $a_n = \sum_{i\geq 0} s(n,i)b_i$. Let $A(x)$ and $B(x)$ be the exponential generating functions of sequences $\{a_n\}_{n=0}^{\infty}$ and $\{b_n\}_{n=0}^{\infty}$, respectively. Obviously $B(x) = \frac{x^k}{k!}$. Lemma 3.3.9 implies that the exponential generating function of $\{s(n,k)\}_{n\geq 0}$ is

$$
A(x) = B(\ln(1+x)) = \frac{(\ln(1+x))^k}{k!}.
$$

$\square$

Definition 3.3.10. *The nth Bell number B_n is the number of all partitions on $[n]$.*

By definition, $B_0 = S(0,0) = 1$, $B_1 = S(1,1) = 1$, and $B_2 = S(2,1) + S(2,2) = 1 + 1 = 2$. Following Lemma 3.3.8,

$$B_n = \sum_{k=0}^{n} S(n,k) = \sum_{k=0}^{n}\sum_{j=0}^{k} \frac{1}{j!(k-j)!} j^n (-1)^{k-j}$$

$$= \sum_{j=0}^{n} \frac{j^n}{j!} \sum_{k=j}^{n} \frac{1}{(k-j)!} (-1)^{k-j} = \sum_{j=0}^{n} \frac{j^n}{j!} \sum_{i=0}^{n-j} \frac{1}{i!} (-1)^i$$

$$= \sum_{j=0}^{n} \frac{j^n}{j!} \exp_{|n-j}(-1), \tag{3.8}$$

where in general $\exp_{|t}(x)$ is an abbreviation of $\sum_{i=0}^{t} \frac{x^i}{i!}$. (3.8) is a computable formula. Nonetheless, we have another explicit formula for the Bell numbers.

Theorem 3.3.7.

$$B_n = \frac{1}{e} \sum_{k \geq 0} \frac{k^n}{k!}.$$

Proof. In a partition of $\{1, \ldots, n\}$, the element n may be contained in a subset of k elements, $1 \leq k \leq n$. Therefore it is not hard to observe the following recursive relation of the Bell numbers:

$$B_n = \sum_{k=1}^{n} \binom{n-1}{k-1} B_{n-k}, \ (n \geq 1)$$

with $B_0 = 1$ and thus $B_1 = 1$. Consider the exponential generating function of Bell numbers

$$B(x) = \sum_{n \geq 0} \frac{B_n}{n!} x^n. \tag{3.9}$$

Taking derivative on both sides of (3.9), we get

$$\frac{dB(x)}{dx} = \sum_{n \geq 1} \frac{B_n}{(n-1)!} x^{n-1}$$

$$= \sum_{n \geq 1} \frac{1}{(n-1)!} \left(\sum_{k=1}^{n} \binom{n-1}{k-1} B_{n-k} \right) x^{n-1}$$

$$= \sum_{n \geq 1} \sum_{k=1}^{n} \frac{x^{k-1}}{(k-1)!} \frac{B_{n-k} x^{n-k}}{(n-k)!}$$

$$= \sum_{k \geq 1} \frac{x^{k-1}}{(k-1)!} \sum_{n \geq k} \frac{B_{n-k} x^{n-k}}{(n-k)!}$$

$$= \sum_{k \geq 1} \frac{x^{k-1}}{(k-1)!} \sum_{i \geq 0} \frac{B_i x^i}{i!}$$

$$= e^x B(x).$$

Solving the differential equation $\frac{dB(x)}{dx} = e^x B(x)$ with initial condition $B(0) = 1$ gives us $B(x) = e^{e^x - 1}$.

Consequently,

$$B(x) = \frac{1}{e} \sum_{k \geq 0} \frac{(e^x)^k}{k!} = \frac{1}{e} \sum_{k \geq 0} \frac{1}{k!} \sum_{n \geq 0} \frac{k^n x^n}{n!}$$

$$= \frac{1}{e} \sum_{n \geq 0} \left(\sum_{k \geq 0} \frac{k^n}{k!} \right) \frac{x^n}{n!}.$$

So,

$$B_n = \frac{1}{e} \sum_{k \geq 0} \frac{k^n}{k!}.$$

□

In [Kla03], several interesting aspects of Bell numbers are investigated. For instance, it is proved that $B(x)$ satisfies no algebraic differential equation over the field of rational functions with complex coefficients.

3.3.2 Harmonic Numbers

Definition 3.3.11. *Arising from truncation of the Harmonic series, the nth Harmonic number H_n is the sum of the reciprocals of the first n positive integers.*

$$H_n = \frac{1}{1} + \frac{1}{2} + \cdots + \frac{1}{n} = \sum_{k=1}^{n} \frac{1}{k}.$$

The numerators of the Harmonic numbers H_n ($n \geq 1$) expressed in irreducible fractions are $1, 3, 11, 25, 137, 49, 363, \ldots$ (OEIS: A001008).

We adopt the usual convention $H_0 = 0$. Let $H(x)$ be the ordinary generating function of H_n:

$$H(x) = \sum_{n \geq 0} H_n x^n.$$

Since $H_n = \sum_{k=1}^{n} \frac{1}{k}$ by definition, $H(x)$ may be written into $H(x) = a(x)b(x)$, where $a(x) = \sum_{n \geq 0} a_n x^n$, $b(x) = \sum_{n \geq 0} b_n x^n$, with $a_0 = 0$, $a_n = \frac{1}{n}$ for all $n \geq 1$ and $b_n = 1$ for all $n \geq 0$.

Clearly $b(x) = \frac{1}{1-x}$. Because $a'(x) = \sum_{n \geq 1} x^{n-1} = \frac{1}{1-x}$, we have $a(x) = -\ln(1-x)$. Thus $H(x) = -\frac{1}{1-x} \ln(1-x) = \frac{1}{1-x} \ln\left(\frac{1}{1-x}\right)$.

Meanwhile, by comparing $\frac{1}{k}$ with $\frac{1}{x}$ for either $x \in [k, k+1]$ or $x \in [k-1, k]$, the integrals $\int_1^{n+1} \frac{1}{x} dx$ and $1 + \int_1^{n} \frac{1}{x} dx$ give us the estimation

$$\ln(n+1) < H_n < \ln n + 1.$$

Bertrand's postulate, or the Bertrand-Chebyshev theorem [Tch52], implies that for every $n \geq 2$, the biggest prime number p in $\{1, \ldots, n\}$ is larger than $n/2$. Write H_n in the form

$$H_n = \frac{\sum_{k=1}^{n} \frac{n!}{k}}{n!} = \frac{\frac{n!}{p} + \sum_{\substack{1 \leq k \leq n \\ k \neq p}} \frac{n!}{k}}{n!},$$

then it is easy to see that the biggest prime p in $\{1, \ldots, n\}$ divides the bottom but not the top. So the Harmonic numbers are never integers for $n \geq 2$.

The Harmonic numbers occur frequently in the fields of number theory and analysis. For instance, write H_n into $H_n = \frac{c_n}{d_n}$ such that c_n and d_n are relatively prime positive integers unless $c_n = 0$ (in which case $n = 0$ and $d_n = 1$). For a prime p, let $J_p = \{n \in \mathbb{N} | p \text{ divides } c_n\}$. Thus $J_2 = \{0\}$, $J_3 = \{0, 2, 7, 22\}$, $J_5 = \{0, 4, 20, 24\}$, etc. For $x \in \mathbb{R}$, let $J_p(x)$ be the number of integers bounded by $[1, x]$. It is conjectured that J_p is finite for all primes p. In [San16], an upper bound of $|J_p(x)|$ is given $(129 p^{0.4} x^{0.765})$. Furthermore, it is shown that for $p > 3$, $\{0, p - 1, p(p - 1), p^2 - 1\} \subseteq J_p$; when equality holds, p is called Harmonic. The first few Harmonic primes are $5, 13, 17, 23, 41$; it is conjectured that there are infinitely many. Necessary and sufficient conditions for p to be Harmonic are given [EL91].

The q-Harmonic numbers of first type and second type, respectively, are defined as follows.

$$H_{[n]} := \sum_{i=1}^{n} \frac{1}{[i]}, \quad H_{\widetilde{[n]}} := \sum_{i=1}^{n} \frac{q^i}{[i]},$$

where $H_{[0]} = H_{\widetilde{[0]}} = 0$; see, e.g., Dilcher [Dil08].

In [Pro08], the following identity involving Harmonic numbers is proved,

$$\sum_{k=0}^{n} (-1)^{n-k} \binom{n+k}{k} \binom{n}{k} H_{n+k} = 2 H_n.$$

In [MSS12], a q-analogue is shown by making use of q-partial fractions.

Theorem 3.3.8. *[MSS12, Theorem 3.6] For all $n \geq 1$,*

$$\sum_{k=1}^{n} (-1)^{n-k} q^{\binom{k-n}{2}} \begin{bmatrix} n+k \\ k \end{bmatrix} \begin{bmatrix} n \\ k \end{bmatrix} H_{[k]} = \sum_{j=1}^{n} \frac{q^{j^2} [2j]}{[j]^2}.$$

The ordinary generalized Harmonic numbers are defined by

$$H_{n,r} = \frac{1}{1^r} + \frac{1}{2^r} + \cdots + \frac{1}{n^r} = \sum_{k=1}^{n} \frac{1}{k^r}.$$

So $H_{n,1}$ is the classical Harmonic number H_n. If $r > 1$, $H_{n,r}$ converges to the *Riemann zeta function* $\zeta(r)$.

There are several types of special numbers that are of significant interests too. These include the Bernoulli numbers, the Eulerian numbers (see Section 2.2.2), the Fibonacci numbers, etc. A wonderful source about these sequences is [GKP89, Chap. 6].

3.4 INTEGER PARTITIONS, TABLEAUX

3.4.1 Integer Partition Numbers

Here we discuss briefly the partition numbers of integers. An integer partition is a way of writing an integer n as a sum of positive integers where the order of the addends is not substantial. By convention, partitions are normally written from largest to smallest addends. The number of integer partitions of a special integer n is denoted by $p(n)$. To be concise, from now on we say "the number of partitions of n" instead of "the number of integer partitions of n" and we call the addends "parts". The number of partitions of n into exactly k parts is denoted by $p(n, k)$. Thus $p(n) = \sum_{k=1}^{n} p(n, k)$. For example, since $5 = 5 = 4+1 = 3+2 = 3+1+1 = 2+2+1 = 2+1+1+1 = 1+1+1+1+1$, we have $p(5) = 7$ and $p(5, 3) = 2$.

Without loss of generality, we set $p(0) = p(0, k) = 0$ for any positive integer k. It is easy to verify the recurrence

$$p(n, k) = p(n - 1, k - 1) + p(n - k, k),$$

which provides a convenient method for making a table of the numbers $p(n, k)$ for small values of n and k.

In 1919, Ramanujan discovered the following three congruences for $p(n)$,

$$p(5n + 4) \equiv 0 \ (\text{mod } 5),$$
$$p(7n + 5) \equiv 0 \ (\text{mod } 7),$$
$$p(11n + 6) \equiv 0 \ (\text{mod } 11),$$

and proclaimed that "it appears that there are no equally simple properties for any moduli involving primes other than these three (i.e. $m = 5, 7, 11$)". [Ram00, p. xix] Ramanujan presented proofs of the first two congruences, and a delicate proof of the third one was given by Winquist [Win69]. Paule and Radu established a recurrence for the generating function of $p(11n + 6)$ [PR17], thus rewitnessed the third congruence. Making a big step from [KO92, Theorem], Ahlgren and Boylan brought closure to the long-standing conjecture of Ramanujan that congruences of the form $p(mn + b) \equiv 0 \ (\text{mod } m)$ are only those three originally observed: when $(m, b) \in \{(5, 4), (7, 5), (11, 6)\}$ [AB03].

Originated from Euler's time, investigation of the deeper properties of $p(n)$ was one of the jewels of 20th century analysis, involving studies of great mathematicians including Hardy, Ramanujan, Rademacher, and later people. This is yet just the beginning of a longer and still ongoing story. In this subsection we introduce only some basic ideas and properties. As we see several times in topics discussed, formal power series and generating function play signal roles in our investigation.

Theorem 3.4.1. *For any $n \in \mathbb{Z}^+$, the number of partitions of n into only odd parts is equal to the number of partitions of n into all distinct parts.*

Proof. We denote the number of partitions of n into only odd parts by $p_o(n)$ and the number of partitions of n into all distinct parts by $p_d(n)$, respectively. Note that the ordinary generating functions of the sequences $\{p_o(n)\}_{n \geq 0}$ and $\{p_d(n)\}_{n \geq 0}$ are respectively

$$\prod_{i=1}^{\infty} \frac{1}{1 - x^{2i-1}}$$

and

$$\prod_{i=1}^{\infty}(1 + x^i).$$

Moreover,

$$\prod_{i=1}^{\infty}(1 + x^i) = \prod_{i=1}^{\infty} \frac{1 - x^{2i}}{1 - x^i} = \frac{\prod_{i=1}^{\infty}(1 - x^{2i})}{\prod_{i=1}^{\infty}(1 - x^i)}$$

$$= \frac{1}{\prod_{i=1}^{\infty}(1 - x^{2i-1})} = \prod_{i=1}^{\infty} \frac{1}{1 - x^{2i-1}}.$$

$\square$

Theorem 3.4.1 was discovered by Euler. A great many proofs of it have been given. Some of the most interesting proofs are bijective. In particular, one due to Glaisher [Gla07] is perhaps the neatest. The reader is recommended to read the artful presentation in [AZ18, Chap. 33].

Theorem 3.4.2. *For any nonnegative integers n and q, the number of partitions of n in which no part appears more than q times is equal to the number of partitions of n into parts that are not divisible by $q + 1$.*

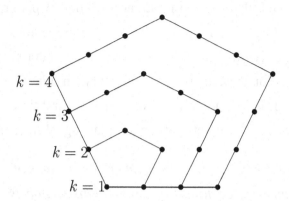

Figure 3.1 The kth pentagonal number is the total number of dots of the kth pentagon.

Proof. On the right side of the following formal power series equality,

$$\prod_{i\geq 1}(1+x^i+\cdots+x^{qi}) = \prod_{i\geq 1}\frac{1-x^{(q+1)i}}{1-x^i},$$

each factor in the numerator of the product cancels one in the denominator, leaving in the denominator only those factors in which i is not divisible by $q+1$. Now it is clear that the two sides are generating functions of the desired sequences, respectively. □

A *pentagonal number* is a polygonal number of the form $\frac{k(3k-1)}{2}$. The first few are $1, 5, 12, 22, 35, 51, 70$ (OEIS A000326). See Figure 3.1 for the combinatorial interpretation of pentagonal numbers. Now let $p_{e,d}(n)$ (resp. $p_{o,d}(n)$) denote the number of partitions of n into an even (resp. odd) number of all *distinct* parts. Amazingly, $p_{e,d}(n) = p_{o,d}(n)$ for virtually all positive integers n except for the case n is either a pentagonal number $\frac{k(3k-1)}{2}$ or its conjugate $\frac{k(3k+1)}{2}$.

Theorem 3.4.3. *For all positive integers n, we have*

$$p_{e,d}(n) - p_{o,d}(n) = \begin{cases} (-1)^k, & \text{if } n = \frac{k(3k\pm 1)}{2}; \\ 0, & \text{otherwise.} \end{cases}$$

Very nice proofs of Theorem 3.4.3 are given by Fabian Franklin (see [And98]), Garsia and Milne [GM81] and other people, among which a bijective construction by David Bressoud and Doron Zeilberger [BZ85] is brilliant. Euler proved his remarkable pentagonal number theorem by calculations with formal series, which became a corollary of Theorem 3.4.3.

Corollary 3.4.1. *[Eul45](Euler's pentagonal number theorem)*

$$\prod_{n \geq 1}(1 - x^n) = 1 + \sum_{k \geq 1}(-1)^k (x^{\frac{3k^2-k}{2}} + x^{\frac{3k^2+k}{2}})$$

$$= \sum_{k=-\infty}^{+\infty}(-1)^k (x^{\frac{3k^2-k}{2}}). \tag{3.10}$$

Proof. The reader is referred to Andrew's marvelous account [And98, p. 11-12]. □

Another effective device for studying integer partitions is graphical representation. For a partition $n = r_1 + \cdots + r_k$ with $r_1 \geq \cdots \geq r_k \geq 1$, we write $\lambda = (r_1, \ldots, r_k) \vdash n$ or $|\lambda| = n$ to mean λ is a partition of n. If the partition λ has α_i parts equal to i, then we write $\lambda = \langle 1^{\alpha_1}, 2^{\alpha_2}, \ldots \rangle$, where terms with $\alpha_i = 0$ and the superscript $\alpha_i = 1$ may be omitted. For example,

$$\lambda = (5, 4, 4, 2, 1) = \langle 1^1, 2^1, 3^0, 4^2, 5^1 \rangle = \langle 1, 2, 4^2, 5 \rangle \vdash 16.$$

Take $\lambda = (\lambda_1, \ldots, \lambda_k) \vdash n$. The Ferrers diagram, or Ferrers graph, of the partition λ is a left-justified array of n dots with λ_i dots in the ith row. The Ferrers diagram of the partition $\lambda = (5, 4, 4, 2, 1)$ (or in other words, of $16 = 5 + 4 + 4 + 2 + 1$) is illustrated in Figure 3.2.

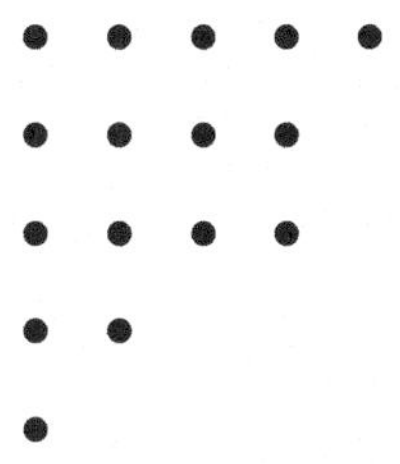

Figure 3.2   Ferrers diagram of the partition $\lambda = (5, 4, 4, 2, 1) \vdash 16$.

By the help of Ferrers diagram representation, one can easily show that

Theorem 3.4.4. *For any integers n and k, the number of partitions of n into exactly k parts is equal to the number of partitions of n with the maximum part k.*

Proof. In fact, we may easily set up a one to one correspondence between the two classes of partitions in consideration by mapping each partition onto its conjugate in terms of the Ferrers diagrams. □

3.4.2 Young Tableaux

An alternative way to represent an integer partition λ is to replace the dots by juxtaposed boxes, and we call the resulting diagram the *Young diagram* of λ. (Sometimes we also call a Young diagram a "Ferrers board".)

Young tableaux were originated by Rev. Alfred Young in his work in invariant theory. Since then, Young tableaux have played important roles in the representation theory of symmetric groups [Ful97, Mac95], in algebraic geometry [Las74, Las75] and in combinatorics [Sta71, Sta99]. In this monograph, we mainly discuss Young tableaux with connections to standard special counting sequences.

Given a partition λ, a semistandard Young tableau (SSYT) of shape λ is an array $T = (T_{ij})$ of positive integers of shape λ, with repetitions allowed, that is weakly increasing in every row and strictly increasing in every column. If the entries of an SSYT T are exactly $1, 2, \ldots, n$, without repetitions, thus strictly increasing in each row and column, T is called a standard Young tableau (SYT). An example of an SYT of shape $(5, 4, 4, 2, 1)$ is given by Figure 3.3.

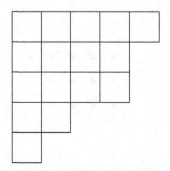

Figure 3.3   Young diagram of the partition $\lambda = (5, 4, 4, 2, 1) \vdash 16$.

1	2	4	9	11
3	5	6	12	
7	8	13	16	
10	15			
14				

Figure 3.4 A standard Young tableau of shape $\lambda = (5, 4, 4, 2, 1) \vdash 16$.

Given a partition λ, we denote by $\mathrm{SYT}(\lambda)$ (resp. $\mathrm{SSYT}(\lambda)$) the set of all standard (resp. semistandard) Young tableaux of shape λ. Furthermore, let

$$f^\lambda := |\mathrm{SYT}(\lambda|,$$

i.e. f^λ denotes the number of SYTs of shape λ.

One of the classical 66 interpretations of the Catalan numbers provided in [Sta99, Ex. 6.19] is the following. The Catalan numbers C_n count the number of SYT of shape (n, n) (or equivalently of shape $(n, n - 1)$): For any fixed $T \in \mathrm{SYT}(n, n)$, simply note that for every $j \in \{1, \dots, 2n\}$, the number of $i \leq j$ in row 1 of T is always as many as the number of $i \leq j$ in row 2 of T, so it is easy to construct a Catalan word w from T; and reversely the Catalan word $w = w_1 w_2 \dots w_{2n} \in CW_n$ uniquely determines $T \in \mathrm{SYT}(n, n)$ by putting $j \in \{1, \dots, 2n\}$ on row 1 if $w_j = 0$ and putting $j \in \{1, \dots, 2n\}$ on row 2 if $w_j = 1$.

By the celebrated Robinson-Schensted-Knuth correspondence or the RSK algorithm, one has

$$\sum_{\lambda \vdash n} (f^\lambda)^2 = n!,$$

as well as one of its consequences,

$$\sum_{\lambda \vdash n} f^\lambda = |\{\sigma \in S_n | \sigma^2 = 1\}| = \sum_{0 \leq i \leq \lfloor n/2 \rfloor} \frac{n!}{2^i i! (n - 2i)!}.$$

(See [Ful97, p. 36-52] or [Sta99, p. 320-331], for instance.)

Fix $T \in \mathrm{SSYT}(\lambda)$. For each cell u of T, the *hook length* at u, denoted $h(u)$, is the number of cells to the right of u or directly below u, counting u itself once. (Thus $h(u)$ is only concerned with the shape λ and the position of the cell u, but irrelevant to the entries T_{ij}.) To be concise, it is often convenient to identify the diagram $\{(i, j)|1 \leq j \leq \lambda_i\}$ with its shape $\lambda = (\lambda_1, \ldots, \lambda_k)$.

9	7	5	4	1
7	5	3	2	
6	4	2	1	
3	1			
1				

Figure 3.5   Hook lengths of the Young tableau shown in Figure 3.4.

The most well-known result on deciding the number of standard Young tableaux of a fixed shape is the *hook length formula*.

Theorem 3.4.5. *The "hook length formula". Given any Young diagram λ of n cells (here we are identifying the diagram $\{(i, j)|1 \leq j \leq \lambda_i\}$ with its shape), we have*

$$f^\lambda = \frac{n!}{\prod_{u \in \lambda} h(u)}.$$

That is, the number of standard Young tableaux with shape λ is exactly n! divided by the product of the hook lengths of all the cells in the diagram of λ.

In [Ful97, p. 54], an intuitive idea to understand Theorem 3.4.5 is mentioned: "There is a quick way to "see" (and remember) this remarkable formula. Consider all n ways to number the n cells of the diagram with the integers from 1 up to n. A numbering will be a tableau exactly when, in each hook, the corner cell of the hook is the smallest among the entries in the hook. The probability of this happening is $1/h$, where h is the hook length. If these probabilities were independent (which they certainly are not!), then the proportion of tableaux among all numberings would be 1 over the product of the hook lengths, and this is the assertion of the proposition. Greene, Nijenhuis, and Wilf (1979) have given a short probabilistic proof of the hook length formula that comes close to justifying this heuristic argument; this proof is also given in Sagan (1991)".

Theorem 3.4.5 is due to Frame, Robinson, and Thrall [FRT54], but for proofs see [GNW79] or [Sag91]. With Theorem 3.4.5, the exercise mentioned earlier that the Catalan numbers C_n count the number of SYT of shape (n, n) [Sta99, Ex. 6.19.ww] becomes an immediate consequence.

6	5	4	3	2
5	4	3	2	1

Figure 3.6   Hook lengths of a Young tableau with shape $\lambda = (n, n)$ ($n = 5$).

A *descent* of T is any instance of i followed by an $i+1$ in a lower row of T, and define the *descent set* $Des(T)$ to be the set of all descents of T. The *major* index of T is defined by $\mathrm{maj}(T) = \sum\limits_{i \in Des(T)} i$.

For $\lambda \vdash n$, set

$$b(\lambda) = \sum_i (i - 1)\lambda_i.$$

Note that $b(\lambda)$ is the smallest possible sum of the entries of an SYT of shape λ, allowing 0 as entries. The sum of $q^{\mathrm{maj}(T)}$ for all $T \in \mathrm{SYT}(\lambda)$ is given by the following q-hook length formula.

Theorem 3.4.6. *The "q-hook length formula". [Sta99, 7.21.5] Given any Young diagram λ of n cells, we have*

$$\sum_{T \in \mathrm{SYT}(\lambda)} q^{\mathrm{maj}(T)} = \frac{q^{b(\lambda)}[n]!}{\prod_{u \in \lambda}[h(u)]}.$$

For proof of Theorem 3.4.6, see [Sta99, p. 304-376]. Like the $q = 1$ case, this provides a q-analogue of the Catalan number when $\lambda = (n, n)$ is a partition with two rows of the same length:

$$\sum_{T \in \mathrm{SYT}(\lambda=(n,n))} q^{\mathrm{maj}(T)} = \frac{q^n}{[n+1]} \begin{bmatrix} 2n \\ n \end{bmatrix}. \tag{3.11}$$

(Be reminded that the diagram $\lambda = (n, n)$ has $2n$ cells now.)

On another direction, (3.11) may be extended to larger sets. An *increasing tableau* is an SSYT such that both rows and columns are strictly increasing, and the set of entries is an initial segment of $\mathbb{Z}^+$ (namely if an integer i appears, then all positive integers less than i should appear). We denote by $\text{Inc}_k(\lambda)$ the set of increasing tableaux of shape λ whose entries are $\{1, 2, \ldots, |\lambda| - k\}$. Thus, $\text{Inc}_0(\lambda) = \text{SYT}(\lambda)$. Besides, if λ has only two rows or two columns, $\text{Inc}_k(\lambda)$ will be the set of increasing tableaux of shape λ, with exactly k numbers appearing twice in its cells.

Pechenik studied the set of $2n$ increasing tableau and obtained the following generalization of (3.11).

Theorem 3.4.7. *[Pec14] For any positive integer n and nonnegative integer k with $k \leq n$, we have*

$$\sum_{T \in \text{Inc}_k(n,n)} q^{\text{maj}(T)} = \frac{q^{n + \binom{k}{2}}}{[n+1]} \begin{bmatrix} n-1 \\ k \end{bmatrix} \begin{bmatrix} 2n-k \\ n \end{bmatrix}. \tag{3.12}$$

Clearly, (3.12) reduces to (3.11) when $n = 0$. In addition, setting $q = 1$, (3.12) indicates that the cardinality of $\text{Inc}_k(n,n)$ is

$$\frac{1}{n+1} \binom{n-1}{k} \binom{2n-k}{n},$$

which is sequence A126216 in [Slo] and is considered a refinement of the little Schröder number $s_n = \frac{S_n}{2}$.

Du et al. introduced and investigated the *row-increasing tableaux*. In [DFZ19], a row-increasing tableau is defined as an SSYT with strictly increasing rows and weakly increasing columns, and the set of entries is a consecutive segment of positive integers. (Note that as an SSYT is supposed to be weakly increasing in rows and strictly increasing in columns, the above definition is due to technical benefit, and a row-increasing tableau is essentially an SSYT with the additional requirement that the set of entries is a consecutive segment of positive integers. For our convenience, we will adopt the ready-to-use notation in [DFZ19] in the account below.)

Given positive integer n, nonnegative integers k and m, and $\lambda \vdash n$, RInc_k^m is the set of row-increasing tableaux of shape λ with the set of entries $\{m + 1, m + 2, \ldots, m + n - k\}$. For simplicity, RInc_k^0 is often written to be RInc_k. Thus for the same λ, $\text{Inc}_k(\lambda) \subseteq \text{RInc}_k(\lambda) \subseteq \text{SSYT}(\lambda)$ and $\text{Inc}_0(\lambda) = \text{RInc}_0(\lambda) = \text{SYT}(\lambda)$.

Theorem 3.4.8. *[DFZ19] For any positive integer n and nonnegative integer k with $k \leq n$, we have*

$$\sum_{T \in \mathrm{RInc}_k(n,n)} q^{\mathrm{maj}(T)} = \frac{q^{n+k(k-3)/2}}{[n-k+1]} \begin{bmatrix} 2n-k \\ k \end{bmatrix} \begin{bmatrix} 2n-2k \\ n-k \end{bmatrix}. \tag{3.13}$$

The reader please check [DFZ19] for a proof of Theorem 3.4.8. If $k = 0$, (3.11) is rediscovered from (3.13). On another hand, specializing $q = 1$ in (3.13) gives

$$|\mathrm{RInc}_k(n,n)| = \frac{1}{n-k+1} \binom{2n-k}{k} \binom{2n-2k}{n-k}. \tag{3.14}$$

The right hand side of (3.14) is a refinement of the (large) Schröder number and is recorded in [Slo] (OEIS: A006318) as well.

CHAPTER 4

Lattice Paths

Many special counting sequences are best interpreted via venue of lattice paths. Studies of enumeration theory of lattice path have been vibrant and rich during the past decades. Among many excellent sources, see for instance, [Moh79], [Nar79], [Ges80], [Sul82], [GV85], [Kra95], [KS99], [Sul00], [Aig01], [KSY07], [EF08], [BMM10], [KP10], [KdM13], , [BKR17], [CU19], [AK19a], [AK19b] and [BKK$^+$19]. Especially, the chapter by Krattenthaler [Kra15] in the *Handbook of Enumerative Combinatorics* by Bóna [B15], is an excellent survey to the literature.

4.1   DYCK PATHS AND PERMUTATION PATHS

4.1.1   Dyck Paths

Definition 4.1.1. *A Dyck path of order n is a lattice path from $(0,0)$ to (n,n) that never goes below the main diagonal $\{(i,i), 0 \le i \le n\}$, with steps $(0,1)$ (or North, for brevity N) and $(1,0)$ (or East, for brevity E). Let $\mathcal{D}_n$ denote the set of all Dyck paths of order n.*

The first natural and important statistic on $\mathcal{D}_n$ is area. Given $\Pi \in \mathcal{D}_n$, area(Π) is defined to be the number of complete squares between Π and the main diagonal line $y = x$. More specifically, let $a_i(\Pi)$ be the number of complete squares in the ith row, from top to bottom, that are below Π and above the main diagonal. The number $a_i(\Pi)$ is also called the *length* of the ith row of Π, and $(a_1(\Pi), a_2(\Pi), \ldots, a_n\Pi)$ is the *area vector* of Π. Finally, area$(\Pi) = \sum_{i=1}^{n} a_i(\Pi)$. An example of a Dyck path of order 6 with area vector $(1,0,0,1,1,0)$ is illustrated in Figure 4.1.

DOI: 10.1201/9781003509912-4

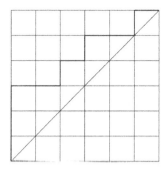

Figure 4.1   A Dyck path $\Pi \in \mathcal{D}_6$ with area$(\Pi) = 5$.

Recall that in Chapter 3, we have introduced Carlitz and Riordan's q-Catalan numbers

$$C_n(q) = \sum_{w \in CW_n} q^{\mathrm{inv}(w)}.$$

Given a Dyck path Π, if we encode each N step by a 0, and each E step by a 1, then from $(0,0)$ to (n,n) we obtain a word $w(\Pi)$ of n 0's and n 1's, which is just a Catalan word. Hence we may associate with each Π the statistics of inv and maj by setting $\mathrm{inv}(\Pi) = \mathrm{inv}(w(\Pi))$ and $\mathrm{maj}(\Pi) = \mathrm{maj}(w(\Pi))$. It is easy to see that $\binom{n}{2} - \mathrm{inv}(\Pi) = \mathrm{area}(\Pi)$. Hence Carlitz and Riordan's q-Catalan numbers $C_n(q)$ may be readily rewritten as

$$C_n(q) = \sum_{\Pi \in \mathcal{D}_n} q^{\binom{n}{2} - \mathrm{area}(\Pi)},$$

or equivalently,

$$C_n(1/q) = q^{-\binom{n}{2}} \sum_{\Pi \in \mathcal{D}_n} q^{\mathrm{area}(\Pi)}.$$

For convenience, define

$$\tilde{C}_n(q) = q^{\binom{n}{2}} C_n(1/q) = \sum_{\Pi \in \mathcal{D}_n} q^{\mathrm{area}(\Pi)}. \tag{4.1}$$

Because there is no essential difference, we also call the modified version $\tilde{C}_n(q)$ Carlitz and Riordan's q-Catalan polynomial.

The following is essentially the same as Theorem 3.1.4 except in different languages.

Theorem 4.1.1.

$$\sum_{\Pi \in \mathcal{D}_n} q^{\text{maj}(\Pi)} = \frac{1}{[n+1]} \begin{bmatrix} 2n \\ n \end{bmatrix}.$$

Over $\mathcal{D}_n$, the statistic bounce was introduced by Haglund in [Hag03]. Here we adopt the description of [HL05] to define it: start by placing a ball at the upper corner (n, n) of a Dyck path Π, then push the ball straight left. Once the ball intersects a vertical step of the path, it "ricochets" straight down until it intersects the diagonal, after which the process is iterated; the ball goes left until it hits another vertical step of the path, then follows down to the diagonal, etc. On the way from (n, n) to $(0, 0)$ the ball will strike the diagonal at various points (i_j, i_j). We define bounce(Π) to be the sum of these i_j. For convenience, we also let the Dyck path so obtained in this process be the *bounce path* of Π and denote it by $b(\Pi)$. In addition, we say Π is *balanced* if and only if $\Pi = b(\Pi)$. In Figure 4.2, a Dyck path Π is represented by the dark line and its bounce path $b(\Pi) = B$ is the pale line. As illustrated, bounce(Π) = 3 + 5 = 8.

In [GH96], Garsia and Haiman introduced a complicated rational function $C_n(q, t)$ which they proved to have the following properties:

$$C_n(q, 1) = \sum_{\Pi \in \mathcal{D}_n} q^{area(\Pi)} = \tilde{C}_n(q),$$

$$q^{\binom{n}{2}} C_n(q, 1/q) = \frac{1}{[n+1]} \begin{bmatrix} 2n \\ n \end{bmatrix}.$$

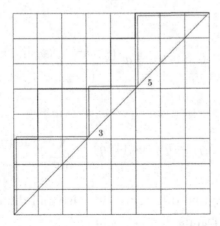

Figure 4.2   A Dyck path Π and its bounce path in pale line.

In order to interpret $C_n(q, t)$, Haglund [Hag03] introduced the distribution function

$$F_n(q, t) = \sum_{\Pi \in \mathcal{D}_n} q^{\text{area}(\Pi)} t^{\text{bounce}(\Pi)}$$

and conjectured that $F_n(q, t) = C_n(q, t)$. Garsia and Haglund ([GH02], [GH01]) proved this by using symmetric function methods, and as a byproduct also the conjecture in [GH96] that $C_n(q, t)$ is a polynomial with positive integer coefficients. Therefore, $C_n(q, t)$ is now called the q, t-*Catalan polynomial*.

4.1.2 Permutation Paths and a Weight-Preserving Bijection between $\mathfrak{S}_n$ and $\mathcal{P}_n$

As an attempt to extend the idea of Dyck paths, we introduce the notion of Permutation paths and develop the related theory.

Definition 4.1.2. *A Permutation path of order n is a lattice path from $(0, 0)$ to (n, n) which never goes below the main diagonal (i, i), $0 \leq i \leq n$, or above the line $y = n$, and consists of North $(0, 1)$, East $(1, 0)$ and SOUTH $(0, -1)$ steps but never repeats (i.e. no North step followed or preceded by an immediate SOUTH step).*

Let $\mathcal{P}_n$ denote the collection of Permutation paths of order n. Figure 4.3 is an illustration of the 6 members in $\mathcal{P}_3$.

Theorem 4.1.2.

$$|\mathcal{P}_n| = n! .$$

Proof. Note that any Permutation path $\Pi \in \mathcal{P}_n$ consists of some North steps, some SOUTH steps and exactly n East steps made at different columns. In more details, for j from 1 to n, these n East steps are in the form of $(j - 1, h_j) \to (j, h_j)$, where h_j could be any integer satisfying $j \leq h_j \leq n$ because Π never goes below the main diagonal or above the line $y = n$. In fact Π is uniquely decided by these East steps or equivalently the sequence of their heights (y-values) $(h_1, \ldots, h_n)$. Once these East steps are fixed, we just connect them up by consecutive North or SOUTH steps, or possibly an empty vertical move if $h_j = h_{j+1}$. Since repeats are not allowed, the connection is unique. Therefore the number of Permutation

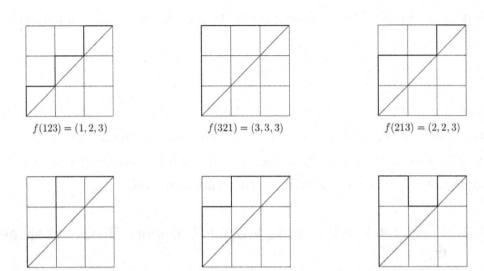

$f(123) = (1, 2, 3)$ $f(321) = (3, 3, 3)$ $f(213) = (2, 2, 3)$

$f(132) = (1, 3, 3)$ $f(231) = (2, 3, 3)$ $f(312) = (3, 2, 3)$

Figure 4.3   Correspondence between $\mathcal{P}_3$ and $\mathfrak{S}_3$ under f.

paths of order n is equal to the number of sequences $(h_1, \ldots, h_n)$. For each j, since $j \leq h_j \leq n$, there are $n + 1 - j$ ways to choose h_j. Furthermore, the choices of different h_j's are independent. So we are done. □

The cardinality of $n!$ naturally motivates us to give a bijection between $\mathcal{P}_n$ and the symmetric group $\mathfrak{S}_n$. Actually the proof of Theorem 4.1.2 already sheds lights on this bijection.

Lemma 4.1.3. *There exists a bijection f between the symmetric group $\mathfrak{S}_n$ and the Permutation paths $\mathcal{P}_n$.*

Proof. For any $\Pi \in \mathcal{P}_n$, define the *height sequence* of Π to be the sequence of the heights (y-values) of Π's East steps, from left to right, as in the proof of Theorem 4.1.2. Denote the height sequence by $h(\Pi) = (h_1^{\Pi}, \ldots, h_n^{\Pi})$. Clearly $j \leq h_j^{\Pi} \leq n$ for each j and any integer vector satisfying this requirement is a height sequence for some uniquely decided Permutation path.

Given $\sigma = \sigma_1 \cdots \sigma_n \in \mathfrak{S}_n$, find its "lifted word" $l(\sigma) = l_1^{\sigma} \cdots l_n^{\sigma}$ where for $1 \leq j \leq n$, l_j^{σ} is what σ_j would become if we map $\{\sigma_j, \ldots, \sigma_n\}$ to the set $\{j, \ldots, n\}$ keeping the relative order of each element.

For example, if $n = 6$ and $\sigma = 6\ 2\ 4\ 3\ 5\ 1$, then $l(\sigma) = 6\ 3\ 5\ 5\ 6\ 6$: $l_1^\sigma = 6$ because $\sigma_1 = 6$ is the biggest among $\{6, 2, 4, 3, 5, 1\}$ when the set $\{6, 2, 4, 3, 5, 1\}$ is mapped to $\{1, 2, 3, 4, 5, 6\}$ where 6 is also the biggest; $l_2^\sigma = 3$ because $\sigma_2 = 2$ is the second smallest in $\{2, 4, 3, 5, 1\}$ when the set $\{2, 4, 3, 5, 1\}$ is mapped to $\{2, 3, 4, 5, 6\}$ where the second smallest element is 3, etc. Notice that l_1^σ is always equal to σ_1, l_n^σ is always n and that $i \le l_i^\sigma \le n$ for every i.

So, $l(\sigma)$ is a height sequence. Find its corresponding Permutation path Π and let $f(\sigma) = \Pi$.

Conversely, given any Permutation path Π, locate its height sequence $h(\Pi)$. Actually we will use $h(\Pi)$ as the "lifted word" to find $\sigma = f^{-1}(\Pi)$. Let $\sigma_1 = h_1^\Pi$. For i from 2 to n, let σ_i be the $(h_i^\Pi + 1 - i)$th smallest number in the set $\{1, \dots, n\} - \{\sigma_1, \dots, \sigma_{i-1}\}$. Clearly $l(\sigma) = h(\Pi)$ and hence the permutation σ so obtained is $f^{-1}(\Pi)$.

So we have established the bijection f as desired. □

Sometimes we use the height sequence to represent a Permutation path for convenience. That is, we may write Π directly as $\Pi = (h_1^\Pi, \dots, h_n^\Pi)$, where $h(\Pi)$ is the height sequence of Π.

Example 4.1.4. *When $n = 3$, there are $3! = 6$ Permutation paths, and their correspondence with the permutations in $\mathfrak{S}_3$ through f is indicated in Figure 4.3. For each Permutation path, the left side is $f(\sigma)$, and the right side is the height sequence.*

The area statistic, previously defined on Dyck paths $\mathcal{D}_n$, may be extended to $\mathcal{P}_n$ naturally since a Permutation path never goes below the main diagonal. Simply, we let $\text{area}(\Pi)$ be the number of complete squares between Π and the main diagonal line $y = x$. This agrees with the definition of area on $\mathcal{D}_n$, which is a subset of $\mathcal{P}_n$, if the Permutation path Π is also a Dyck path.

The bijection f has the nice property of mapping *inv* to area.

Theorem 4.1.3. *f is a weight-preserving bijection between $\mathfrak{S}_n$ and $\mathcal{P}_n$ that maps the inversion statistic to the area statistic. Namely, for any $\sigma \in \mathfrak{S}_n$, we have*

$$\text{inv}(\sigma) = \text{area}(f(\sigma)).$$

Proof. Let $f(\sigma) = \Pi$ and $h(\Pi) = (h_1^\Pi, \ldots, h_n^\Pi)$. Notice that

$$\text{area}(\Pi) = \sum_{i=1}^{n} h_i^\Pi - i$$

and for $1 \leq i \leq n - 1$ $(h_n^\Pi - n = 0)$ we have

$$h_i^\Pi - i = |\{j : \sigma_i > \sigma_j \text{ and } i < j \leq n\}|.$$

So it is clear. □

Corollary 4.1.5.

$$\sum_{\Pi \in \mathcal{P}_n} q^{\text{area}(\Pi)} = [n]!.$$

Proof. Recall that the inv statistic is Mahonian on $\mathfrak{S}_n$ (Theorem 2.2.1),

$$\sum_{\sigma \in \mathfrak{S}_n} q^{\text{inv}(\sigma)} = [n]!.$$

So the conclusion follows from Theorem 4.1.3. Alternatively, it could also be proved by induction. □

As we did for Dyck paths, we may associate each Permutation path with an appropriate word. Still encode each N step by a 0, each E step by a 1, and in addition encode each S step by a special character 0^*.

Definition 4.1.6. *A Permutation word of order n is a permutation of the multiset $\{0^{n+s}, 1^n, 0^{*s}\}$, where $1 \leq s \leq \lfloor \frac{n^2}{4} \rfloor$, with the property that for $1 \leq i \leq 2n + 2s$, in the initial subword $w = w_1 w_2 \cdots w_i$,*

- *the number of 0's is at least as many as the sum of the numbers of 1's and the number of 0^*'s;*

- *the number of 0's minus the number of 0^*'s is at most n;*

- *0 and 0^* are never adjacent.*

We let PW_n denote the set of Permutation words of order n.

Definition 4.1.7. *The inversion statistic of a Permutation word* $w = w_1 w_2 \cdots w_{2n+2s}$ *is defined to be*

$$\text{inv}(w) = \sum_{i:\ w_i=1} (n_1(i) - n_2(i)),$$

where $n_1(i)$ *is the number of 0's after* w_i *and* $n_2(i)$ *is the number of* 0^**'s after* w_i.

The inversion statistic so defined on PW_n is an extension of the inversion statistic on the Catalan words CW_n. The following theorem elucidates our purpose of definition.

Theorem 4.1.4.

$$\sum_{w \in PW_n} q^{\binom{n}{2} - \text{inv}(w)} = [n]!.$$

Proof. For any $w \in PW_n$, it is easy to see that

$$\binom{n}{2} - \text{inv}(w) = \text{area}(\Pi(w)),$$

where $\Pi(w)$ is the Permutation path that w corresponds to. $\qquad\square$

4.1.3 Restricting the Map to Special Pattern Forbidding Permutations

Since Dyck paths $\mathcal{D}_n$ is a subset of $\mathcal{P}_n$ and at least the area statistic is extended to $\mathcal{P}_n$ in a natural way, we study $f^{-1}(\mathcal{D}_n)$ and some other related objects with the hope to understand the Catalan phenomena better.

First we need some preliminary background on the theory of patterns.

Given permutations $\tau \in \mathfrak{S}_k$ and $\sigma \in \mathfrak{S}_n$, we define an *occurrence* of the pattern τ in σ to be a choice of k slots

$$1 \le i_1 < \cdots < i_k \le n$$

such that the sequence $\sigma_{i_1}, \ldots, \sigma_{i_k}$ is in the same order of relative size as the sequence $\tau_1, \ldots, \tau_k$. In other words, for $1 \le j_1 < j_2 \le k$,

$$\sigma_{i_{j_1}} < \sigma_{i_{j_2}} \text{ iff } \tau_{j_1} < \tau_{j_2}.$$

Sometimes we also say that $(\sigma_{i_1}, \ldots, \sigma_{i_k})$ is a τ-*occurrence* in σ. (By the language used in 2.3 it says that the reduction of the integer sequence $\sigma_{i_1}\sigma_{i_2}\cdots\sigma_{i_k}$ is τ.)

Accordingly, if σ does not contain any τ-occurrences of the pattern, we say that σ is τ-*avoiding*. Denote the set of all τ-avoiding permutations in $\mathfrak{S}_n$ by $\mathfrak{S}_n(\tau)$ [Pri97].

Example 4.1.8. *Consider* $\sigma = 51324 \in \mathfrak{S}_5$ *and* $\tau = 123 \in \mathfrak{S}_3$. σ *is NOT* τ-*avoiding because*

$$(\sigma_2, \sigma_3, \sigma_5) = (1, 3, 4)$$

is a τ-*occurrence in* σ. *Notice that*

$$(\sigma_2, \sigma_4, \sigma_5) = (1, 2, 4)$$

is also a τ-*occurrence in* σ, *but finding one occurrence is sufficient for our purpose here.*

Alternatively, let's consider $\hat{\sigma} = 32541 \in \mathfrak{S}_5$ *and* $\hat{\tau} = 312 \in \mathfrak{S}_3$. *Then* $\hat{\sigma}$ *is* $\hat{\tau}$-*avoiding because we can not find any 312-occurrence in* $\hat{\sigma} = 32541$. *So we can say that*

$$32541 \in \mathfrak{S}_5(312).$$

The theory of pattern avoidance has been studied extensively. It is now well known (see, for example, [Knu73]) that for any $\tau \in \mathfrak{S}_3$, $|\mathfrak{S}_n(\tau)| = C_n$. Partly because of this, we are motivated to consider $f^{-1}(\mathcal{D}_n)$, and we find that this pre-image is indeed $\mathfrak{S}_n(312)$. In fact, there have been different direct bijections between $\mathfrak{S}_n(312)$ and $\mathcal{D}_n$ in recent literature: [Kra01] and [BK01]. The latter also gives a weight-preserving bijection exchanging the inv and area statistics but it is different from our f.

Let's consider the restriction of f on $\mathfrak{S}_n(312)$, the 312-*avoiding* permutations, and call it f^*. We prove that f^* is a bijective map between $\mathfrak{S}_n(312)$ and Dyck paths $\mathcal{D}_n$, a subset of the image set of Permutation paths.

Theorem 4.1.5. *[Son05b]* f^* *is a weight-preserving bijection between* $\mathfrak{S}_n(312)$ *and Dyck paths* $\mathcal{D}_n$ *that maps the inversion statistic to the area statistic, and therefore*

$$\sum_{\sigma \in \mathfrak{S}_n(312)} q^{\mathrm{inv}(\sigma)} = \sum_{\Pi \in \mathcal{D}_n} q^{\mathrm{area}(\Pi)},$$

where the right hand side is Carlitz and Riordan's q-Catalan polynomial $\tilde{C}_n(q)$ (4.1) *that satisfies the recurrence*

$$\tilde{C}_n(q) = \sum_{k=1}^{n} q^{k-1} \tilde{C}_{k-1}(q) \tilde{C}_{n-k}(q).$$

Proof. Any Permutation path $\Pi \in \mathcal{P}_n$ can be uniquely represented by its height vector $h(\Pi) = (h_1^{\Pi}, \ldots, h_n^{\Pi})$. Notice that $\Pi \in \mathcal{D}_n$ if and only if for $1 \leq i \leq n-1$, so we have

$$h_i^{\Pi} \leq h_{i+1}^{\Pi}.$$

Given $\sigma \in \mathfrak{S}_n(312)$, we prove $f(\sigma) = \Pi \in \mathcal{D}_n$. Recall that $h(\Pi) = l(\sigma)$. This means for $1 \leq i \leq n$, $h_i^{\Pi} = l_i^{\sigma}$ is what σ_i would become if we map $(\sigma_i, \ldots, \sigma_n)$ to the array $(i, \ldots, n)$ keeping the relative order of each element. Revising this a little bit, let $(l_{i+1}^{\sigma})^-$ denote what σ_{i+1} would be at the "previous stage", i.e, when we map $(\sigma_i, \ldots, \sigma_n)$ to $(i, \ldots, n)$ rather than replacing i by $i+1$ (for which we would get l_{i+1}^{σ}). Now note that

- If $\sigma_{i+1} > \sigma_i$, then $l_{i+1}^{\sigma} = (l_{i+1}^{\sigma})^- > l_i^{\sigma}$.

- If $\sigma_{i+1} < \sigma_i$, then $l_{i+1}^{\sigma} = (l_{i+1}^{\sigma})^- + 1$. So

$$l_{i+1}^{\sigma} \geq l_i^{\sigma}$$
$$\Leftrightarrow (l_{i+1}^{\sigma})^- \geq l_i^{\sigma} - 1$$
$$\Leftrightarrow \nexists k, \text{s.t. } i < i+1 < k \text{ and } \sigma_{i+1} < \sigma_k < \sigma_i.$$

From the above conditions, it is clear that $\sigma \in \mathfrak{S}_n(312)$ implies $\Pi \in \mathcal{D}_n$.

Conversely, given $\Pi \in \mathcal{D}_n$, consider its pre-image $f^{-1}(\Pi) = \sigma$. Suppose $\sigma \notin \mathfrak{S}_n(312)$. Take a *minimal* 312-occurrence $\{\sigma_i, \sigma_j, \sigma_k\}$ in the sense

$$i < j < k,$$
$$\sigma_j < \sigma_k < \sigma_i,$$
$$\text{and } |j - i| + |k - i| \text{ is } minimal.$$

$\Pi \in \mathcal{D}_n$ implies that $h_i^{\Pi} \leq h_{i+1}^{\Pi}$, and hence $l_i^{\sigma} \leq l_{i+1}^{\sigma}$. This requires $\sigma_i - \sigma_{i+1} \leq 1$, and hence $j \neq i+1$. Then what about σ_{i+1}? If $\sigma_{i+1} > \sigma_k$, then $(\sigma_{i+1}, \sigma_j, \sigma_k)$

would be another 312-occurrence, violating the minimality. So $\sigma_{i+1} < \sigma_k$. But then $(\sigma_i, \sigma_{i+1}, \sigma_k)$ would be a "less" 312-occurrence. This shows that $\sigma \notin \mathfrak{S}_n(312)$ is impossible.

Therefore $f(\mathfrak{S}_n(312)) = \mathcal{D}_n$, or we may say that there exists a bijection f^* from $\mathfrak{S}_n(312)$ to $\mathcal{D}_n$.

Since f is weight-preserving, its restriction f^* is also weight-preserving. □

The above result gives rise to more general questions. Since we now know $f^{-1}(\mathfrak{S}_n(312)) = \mathcal{D}_n$, what is $f^{-1}(\mathfrak{S}_n(k12\ldots(k-1)))$ for general k? We answer this question partially by giving a lower bound.

For convenience, let the restriction of f on $\mathfrak{S}_n(k12\cdots(k-1))$, the set of $k12\cdots(k-1)$-*avoiding* permutations, be $f^{(k)}$. To explain our combinatorial interpretation, we need the following definition.

Definition 4.1.9. *An m-cave of a Permutation path Π, with height sequence $h(\Pi) = h_1^\Pi h_2^\Pi \cdots h_n^\Pi$, is a step i satisfying that*

$$max\{h_1^\Pi, \ldots, h_{i-1}^\Pi\} - h_i^\Pi = m,$$

where $m \geq 1$. For an m-cave c, sometimes we say that c is of depth m. An m-triangle is a sequence of caves, $c_1, c_2, \ldots, c_m$, not necessarily continuous but in order from left to right, where c_j is of depth at least $m+1-j$ for each $1 \leq j \leq m$. If a Permutation path Π does not contain any m-triangle, we say that Π is m-triangle-forbidding. Denote the set of m-triangle-forbidding Permutation paths of order n by $\mathcal{F}_{n,m}$.

Geometrically, an m-cave of $\Pi \in \mathcal{P}_n$ is a step which is m squares down compared with the highest level that Π has reached earlier (observe that the highest level that Π will reach later is always $y = n$). The juxtaposition of the not-necessarily continuous sequence of caves in an m-triangle contains an isosceles right triangle with leg length m. Since we require $m \geq 1$, a Dyck path has no m-cave or m-triangle. In addition, $\mathcal{F}_{n,1} = \mathcal{D}_n$. The following figure illustrates a path $\Pi \in \mathcal{P}_8$ with three caves: c_1 at column 2 is of depth 1, c_2 at column 5 is of depth 2, and c_3 at column 6 is of depth 1. The cave c_1 itself is a 1-triangle, and so is every other cave. The sequence of caves $\{c_2, c_3\}$ forms the only 2-triangle of Π and there is no m-triangle for $m \geq 3$.

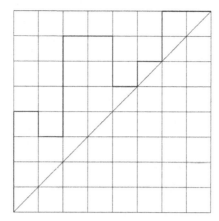

Figure 4.4   A Permutation path $\Pi \in \mathcal{P}_8$ with 3 caves. and a 2-triangle.

The next theorem provides a combinatorial interpretation of the $(m1 \cdots (m-1))$-avoiding permutations in terms of the $(m-2)$-triangle-forbidding paths.

Theorem 4.1.6. *For $m \geq 3$, if $\sigma \in \mathfrak{S}_n$ contains an $m1 \cdots (m-1)$ pattern, then the Permutation path $\Pi = f(\sigma)$ must have an $(m-2)$-triangle. That is,*

$$\mathcal{F}_{n,m-2} \subseteq f(\mathfrak{S}_n(m1 \cdots (m-1))).$$

When $m = 3$, $\subseteq$ is replaced by equality.

Proof. For any $\sigma \notin \mathfrak{S}_n(m12 \cdots (m-1))$, we prove $\Pi = f(\sigma)$ contains some $(m-2)$-triangle. Assume $\{\sigma_{i_1}, \ldots, \sigma_{i_m}\}$ is a $(m12 \cdots m-1)$-occurrence in σ, namely

$$i_1 < i_2 < \cdots < i_m,$$

$$\sigma_{i_2} < \cdots \sigma_{i_m} < \sigma_{i_1}.$$

In fact, for j from 2 to $m-1$, we show that there is a d_j-cave at step i_j, where $d_j \geq m - j$. Note that

$$h_{i_2}^{\Pi} < \cdots < h_{i_{m-1}}^{\Pi}.$$

On top of that, because $\sigma_{i_{m-1}} < \sigma_{i_m} < \sigma_{i_1}$, we have

$$h_{i_{m-1}}^{\Pi} < h_{i_1}^{\Pi}.$$

So,

$$max\{h_1^\Pi, \ldots, h_{i_j-1}^\Pi\} - h_{i_j}^\Pi \geq h_{i_1}^\Pi - h_{i_j}^\Pi \geq m - j.$$

Hence at step $i_j, 2 \leq i_j \leq m - 1$, we have a cave of depth at least $m - j$ and therefore $\Pi = f(\sigma)$ contains an $(m - 2)$-triangle.

$m = 3$ is the case of Dyck paths discussed in Theorem 4.1.5. $\qquad\qquad\square$

Remark 4.1.10. *Conversely, given $\sigma \in \mathfrak{S}_n(m12\ldots m - 1)$, $\Pi = f(\sigma)$ does not necessarily forbid $(m - 2)$-triangles. For example, letting $n = 8$ and $m = 4$, $\sigma = 57836241 \in \mathfrak{S}_8(4123)$, the corresponding Permutation path Π is not 2-triangle-forbidding. See Figure 4.5.*

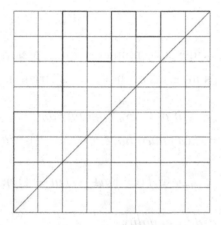

Figure 4.5 $\sigma = 57836241 \in \mathfrak{S}_8(4123)$ but Π is not 2-triangle-forbidding.

Nonetheless, we now know that in terms of cardinality,

$$|\mathcal{F}_{n,m-2}| \leq |\mathfrak{S}_n(m12\ldots m - 1)|.$$

A result of Backelin, West and Xin ([BWX07], see also [SW02]) implies that for any $k \in \mathbb{N}$,

$$|\mathfrak{S}_n(m12\ldots m - 1) = \mathfrak{S}_n(12\ldots m)|.$$

So Theorem 4.1.6 provides a lower bound estimate of the number of permutations in $\mathfrak{S}_n$ whose longest increasing subsequence has length m.

4.2 SCHRÖDER PATHS, ETC, THE GENERALIZED VERSIONS

4.2.1 Schröder Paths

Recall that a d-Schröder word of order n is a word of length $2n + d$, consisted of $n - d$ 0's, d 1's, and $n - d$ 2's, and that the cardinality of all Schröder words of order n is the nth Schröder number S_n.

In the following, we are concerned with the *Schröder paths with d diagonal steps.*

Definition 4.2.1. *A Schröder path of order n and with d diagonal steps is a lattice path from $(0,0)$ to (n,n) that never goes below the main diagonal $\{(i,i), 0 \le i \le n\}$, with $(0,1)$ (or North, for brevity N), $(1,0)$ (or East, for brevity E) and exactly d $(1,1)$ (or Diagonal, for brevity D) steps. Let $\mathcal{S}_{n,d}$ denote the set of all Schröder paths of order n and with d diagonal steps.*

The number of Schröder paths of order n and with d diagonal steps is counted by

$$S_{n,d} = \binom{2n-d}{d} C_n$$

$$= \frac{1}{n-d+1} \binom{2n-d}{d, n-d, n-d}.$$

Clearly $S_n = \sum_{d=0}^{n} S_{n,d}$ and $C_n = S_{n,0}$.

Much of the theory about Dyck paths can be generalized to Schröder paths. In general for a lattice path Π that never goes below the diagonal line $x = y$, define *lower triangle* to be a triangle with vertices (i,j), $(i+1,j)$ and $(i+1,j+1)$, and let the area of Π, denoted by area(Π), be the number of lower triangles between Π and the main diagonal. This new definition of area agrees with the old one for Dyck paths, and is well defined for Schröder paths. On the other hand, if we map $\mathcal{S}_{n,d}$ to the words of $n - d$ 0's, d 1's and $n - d$ 2's by replacing each N step by a 0, each D step by a 1 and each E step by a 2 in a Schröder path Π, then we have the maj statistic for Schröder paths. Bonin, et. al. showed that

Theorem 4.2.1. *[BSS93]*

$$\sum_{\Pi \in \mathcal{S}_{n,d}} q^{\mathrm{maj}(\Pi)} = \frac{1}{[n-d+1]} \begin{bmatrix} 2n-d \\ n-d, n-d, d \end{bmatrix}. \tag{4.2}$$

Remark 4.2.2. *Theorem 4.2.1 is in fact the lattice path version of Theorem 3.2.1. The right side of (4.2) is a desirable and awesome q-Schröder polynomial.*

In Figure 4.6 below, the Schröder path $\Pi \in \mathcal{S}_{8,4}$ is encoded by 000110221212, which implies that $\mathrm{maj}(\Pi) = 5+8+10 = 23$, and has area vector (0,1,2,3,3,3,2,1), which says $\mathrm{area}(\Pi) = 0+1+2+3+3+3+2+1 = 15$. The length of each row, as computed from the number of lower triangles, is shown on the right.

In [EHKK03], Egge, et al. generalized bounce to Schröder paths through a decomposition procedure and defined the following (q, t)-Schröder polynomial

$$S_{n,d}(q,t) = \sum_{\Pi \in \mathcal{S}_{n,d}} q^{\mathrm{area}(\Pi)} t^{\mathrm{bounce}(\Pi)}.$$

They generalized Garsia and Haiman's result to the following

$$q^{\binom{n}{2}-\binom{d}{2}} S_{n,d}\left(q, \frac{1}{q}\right) = \frac{1}{[n-d+1]} \begin{bmatrix} 2n-d \\ n-d, n-d, d \end{bmatrix},$$

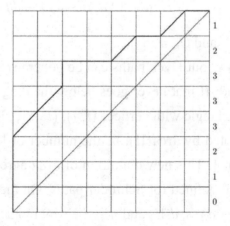

Figure 4.6 A Schröder path $\Pi \in \mathcal{S}_{8,4}$ with $\mathrm{area}(\Pi) = 15$ and $\mathrm{maj}(\Pi) = 23$.

They also conjectured that the (q,t)-Schröder polynomial is symmetric and made a stronger conjectural interpretation of $S_{n,d}(q,t)$ involving a linear operator ∇ defined on the modified Macdonald basis (for details see [EHKK03], [Hag04] or [HL05]).

Theorem 4.2.2. *For all integers n, d with $d \leq n$,*

$$S_{n,d}(q,t) = < \nabla e_n, e_{n-d}h_d > .$$

The above was proved in [Hag04] and thus became the (q,t)-*Schröder Theorem.*

If we remove the requirement "that never goes below the main diagonal" in Definition 4.2.1, then we obtain a central Delannoy path.

Definition 4.2.3. *A central Delannoy path of order n and with d diagonal steps is a lattice path from $(0,0)$ to (n,n) with $(0,1)$, $(1,0)$ and exactly d (1,1) steps. More generally, an (m,n,d)-Delannoy path is a lattice path from $(0,0)$ to (m,n) with $(0,1)$, $(1,0)$ and exactly d (1,1) steps.*

Clearly, the set of (m,n,d)-Delannoy paths is counted by

$$D(m,n,d) = \binom{m+n-d}{m-d, n-d, d}.$$

Moreover, $\sum_{d \geq 0} D(m,n,d)$ is the (m,n)-Delannoy number $D(m,n)$ discussed in the previous chapter, and that the central Delannoy number $D(n,n) = \sum_{d \geq 0} D(n,n,d)$ counts the central Delannoy paths of order n regardless of the number of diagonal steps.

4.2.2 Higher Dimensions

While the standard Catalan and Schröder theories both have been extensively studied, people have only begun to investigate more general cases. There are two major directions. One direction is to replace the original lattice square by a rectangle (see [HPW99] and [GH96]). Another direction is to consider higher dimensional lattice paths in a cube or even a hyper cube. For the latter case, by using a combinatorial cancellation argument and a result of MacMahon [Mac78, p.455-6] [Lot97],

Sulanke [Sul05] derived a formula for 3-Narayana number:

$$N(3, n, k) = \sum_{j=0}^{k} (-1)^{k-j} \binom{3n+1}{k-j} \prod_{i=0}^{2} \binom{n+i+j}{n} \binom{n+i}{n}^{-1},$$

for the number of lattice paths Π using the steps $X := (1, 0, 0)$, $Y := (0, 1, 0)$, and $Z := (0, 0, 1)$, running from $(0, 0, 0)$ to (n, n, n), lying in the chamber $\{(x, y, z) : 0 \leq x \leq y \leq z\}$, with $\mathrm{asc}(\Pi) = k$, where the ascent set $\mathrm{asc}(\Pi)$ is defined to be $\mathrm{asc}(\Pi) = |\{i : \Pi_i \Pi_{i+1} \in \{XY, XZ, YZ\}, 1 \leq i \leq 3n - 1\}|$.

Below we discuss in more details the former direction, i.e., the planar m-Schröder theory. First let's introduce the notions of generalized Dyck and Schröder paths in a rectangle.

Definition 4.2.4. *An m-Dyck path of order n is a lattice path from $(0, 0)$ to (mn, n) which never goes below the main diagonal $\{(mi, i) : 0 \leq i \leq n\}$, with steps $(0, 1)$ (or North, for brevity N) and $(1, 0)$ (or East, for brevity E). Let $\mathcal{D}_n^m$ denote the set of all m-Dyck paths of order n.*

A 2-Dyck path of order 5 is illustrated in Figure 4.7.

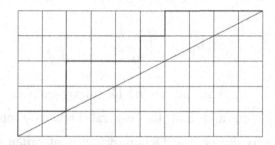

Figure 4.7   A 2-Dyck path in $\mathcal{D}_5^2$.

As in the $m = 1$ case, given $\Pi \in \mathcal{D}_n^m$, we encode each N step by a 0 and each E step by a 1 so as to obtain a word $w(\Pi)$ of n 0's and mn 1's. This clearly provides a bijection between $\mathcal{D}_n^m$ and CW_n^m, where

$$CW_n^m = \left\{ w \in M_{n,mn} \middle| \begin{array}{c} \text{at any initial segment of } w, \text{ the number of 0's times} \\ m \text{ is at least as many as the number of 1's.} \end{array} \right\}$$

We call this special set of $\{0, 1\}$ words, CW_n^m, *Catalan words of order n and dimension m.*

It is shown in [HP91] (see also [HPW99]) that the number of m-Dyck paths, denoted by C_n^m, is equal to

$$\frac{1}{mn+1}\binom{mn+n}{n},$$

which we call the m-Catalan number. In fact, Cigler [Cig87] proved that the number of m-Dyck paths with k peaks, i.e., those with exactly k consecutive NE pairs, is the generalized Narayana number,

$$N_{n,k-1}^m = \frac{1}{n}\binom{n}{k}\binom{mn}{k-1}. \tag{4.3}$$

The read should compare (4.3) with (3.5).

Now we turn to the more general m-Schröder theory.

Definition 4.2.5. *An m-Schröder path of order n is a lattice path from $(0,0)$ to (mn, n) which never goes below the main diagonal $\{(mi, i) : 0 \le i \le n\}$, with steps $(0,1)$ (or North, for brevity N), $(1,0)$ (or East, for brevity E) and (1,1) (or Diagonal, for brevity D). Let $\mathcal{S}_n^m$ denote the set of all m-Schröder paths of order n.*

Definition 4.2.6. *An m-Schröder path of order n with d diagonal steps is a lattice path from $(0,0)$ to (mn, n) which never goes below the main diagonal $\{(mi, i) : 0 \le i \le n\}$, with $(0,1)$ (or North, for brevity N), $(1,0)$ (or East, for brevity E) and exactly d (1,1) (or Diagonal, for brevity D) steps. Let $\mathcal{S}_{n,d}^m$ denote the set of all m-Schröder paths of order n and with d diagonal steps.*

A 2-Schröder path of order 5 and with 3 diagonal steps is illustrated in Figure 4.8.

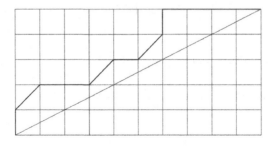

Figure 4.8   A 2-Schröder path in $\mathcal{S}_{5,3}^2$.

Theorem 4.2.3. *[Son05a] The number of m-Schröder paths of order n and with d diagonal steps, denoted by $S_{n,d}^m$, is equal to*

$$\frac{1}{mn-d+1}\binom{mn+n-d}{mn-d,\,n-d,\,d}. \tag{4.4}$$

Proof. For an m-Dyck path Π, let its number of peaks, or consecutive NE pairs, be denoted by $peak(\Pi)$. Notice that any m-Schröder path with d diagonal steps can be obtained uniquely by choosing d of the peaks of a uniquely decided m-Dyck path Π of the same order, and changing each of the chosen consecutive NE pair steps to a Diagonal step. Conversely, given an m-Dyck path Π of order n, choosing d of its peaks (if there are d to choose) and changing them to D steps will give a path in $S_{n,d}^m$. For example, the 2-Schröder path as illustrated in Figure 4.8 is one of $\binom{4}{2} = 6$ paths in $\mathcal{S}_{6,4}^2$ that can be obtained from the 2-Dyck path shown in Figure 4.7. Hence,

$$
\begin{aligned}
S_{n,d}^m &= \sum_{\Pi \in \mathcal{D}_n^m} \binom{peak(\Pi)}{d} \\
&= \sum_{k \geq d} \binom{k}{d} N_{n,k-1}^m \text{ (refer to (4.3))} \\
&= \sum_{k \geq d} \binom{k}{d} \frac{1}{n}\binom{n}{k}\binom{mn}{k-1} \\
&= \frac{\binom{n}{d}}{n} \sum_{k \geq d} \binom{n-d}{n-k}\binom{mn}{k-1} \\
&= \frac{\binom{n}{d}}{n}\binom{mn+n-d}{n-1} \\
&= \frac{1}{mn-d+1}\binom{mn+n-d}{d,\,n-d,\,mn-d}.
\end{aligned}
$$

Above we used the Vandermonde's convolution □

As a generalization of the $m = 1$ case, we name

$$S_n^m = \sum_{d=0}^{n} \frac{1}{mn-d+1}\binom{mn+n-d}{mn-d,\,n-d,\,d}$$

the *m-Schröder number.*

4.2.3 q-m-Schröder Polynomials

When Bonin, Shapiro and Simion [BSS93] studied q-analogues of the Schröder numbers, they obtained several classical results of the single variable cases. Here we generalize some of them to the m case.

Definition 4.2.7. *Define the m-diag polynomial $d_n^m(q)$ over the m-Schröder paths of order n to be*

$$d_n^m(q) = \sum_{\Pi \in \mathcal{S}_n^m} q^{diag(\Pi)},$$

where $diag(\Pi)$ is the number of D steps in the path Π.

Theorem 4.2.4. $d_n^m(q)$ *has $q = -1$ as a root.*

Proof. We use the idea of [BSS93]. The statement is equivalent to say that there are as many m-Schröder paths of order n with an even number of D steps as there are with an odd number of D steps. For any $\Pi \in \mathcal{S}_n^m$, there must be some first occurrence of either a consecutive NE pair of steps, or a D step. According to which occurs first, either replace the consecutive NE pair by a D, or replace the D with a consecutive NE pair. Notice that this presents a bijection between the two sets of objects we wish to show have the same cardinality. □

Recall that (3.3) is a refinement of the q-Catalan identity that says,

$$\sum_{k \geq 1} \sum_{w \in CW_{n,k}} q^{majw} = \sum_{k \geq 1} \frac{1}{[n]} \begin{bmatrix} n \\ k \end{bmatrix} \begin{bmatrix} n \\ k - 1 \end{bmatrix} q^{k(k-1)} = \frac{1}{[n+1]} \begin{bmatrix} 2n \\ n \end{bmatrix}, \quad (4.5)$$

where $CW_{n,k}$ is the set of Catalan words consisting of n 0's, n 1's, with k ascents (i.e. $k - 1$ descents or the corresponding Dyck path has k peaks). As for the m-version, Cigler showed there are exactly $\frac{1}{n} \binom{n}{k} \binom{mn}{k-1}$ m-Dyck paths with k peaks 4.3. In order to generalize the results of (4.5), we derive the following q-identity.

Theorem 4.2.5.

$$\sum_{k \geq d} \begin{bmatrix} k \\ d \end{bmatrix} \frac{1}{[n]} \begin{bmatrix} n \\ k \end{bmatrix} \begin{bmatrix} mn \\ k - 1 \end{bmatrix} q^{(k-d)(k-1)} = \frac{1}{[mn-d+1]} \begin{bmatrix} mn+n-d \\ d, n-d, mn-d \end{bmatrix}.$$

The proof of Theorem 4.2.5 utilizes the q-Vandermonde convolution as introduced in Lemma 2.1.1.

Proof.

$$\sum_{k \geq d} \begin{bmatrix} k \\ d \end{bmatrix} \frac{1}{[n]} \begin{bmatrix} n \\ k \end{bmatrix} \begin{bmatrix} mn \\ k-1 \end{bmatrix} q^{(k-d)(k-1)}$$

$$= \frac{\begin{bmatrix} n \\ d \end{bmatrix}}{[n]} \sum_{k=d}^{n} \begin{bmatrix} n-d \\ n-k \end{bmatrix} \begin{bmatrix} mn \\ k-1 \end{bmatrix} q^{(k-d)(k-1)}$$

$$= \frac{\begin{bmatrix} n \\ d \end{bmatrix}}{[n]} \sum_{j=0}^{n-d} \begin{bmatrix} n-d \\ j \end{bmatrix} \begin{bmatrix} mn \\ n-1-j \end{bmatrix} q^{(n-d-j)(n-1-j)} \text{(by the } q\text{-Vandermonde convolution)}$$

$$= \frac{\begin{bmatrix} n \\ d \end{bmatrix}}{[n]} \begin{bmatrix} mn+n-d \\ n-1 \end{bmatrix}$$

$$= \frac{1}{[mn-d+1]} \begin{bmatrix} mn+n-d \\ d, n-d, mn-d \end{bmatrix}.$$

□

Remark 4.2.8. *It is difficult to find a combinatorial interpretation for the left hand side of Theorem 4.2.5. As a matter of fact, the most straightforward generalization of (4.5) fails even for the 2-Dyck paths:*

$$\sum_{w \in CW_2^2} q^{\text{majw}} = 1 + q^2 + q^3 \neq \frac{[1]}{[5]} \begin{bmatrix} 6 \\ 2 \end{bmatrix} = 1 + q^2 + q^4.$$

Similar to the manner of [HL05], for an m-Dyck path Π of order n, we may associate it with m-*parking functions* by placing one of the n "cars", denoted by the integers 1 through n, in the square immediately to the right of each N step of Π, with the restriction that if car i is placed immediately on top of car j, then $i > j$. Let $\mathbb{P}_n^m$ denote the collection of m-parking functions on n cars.

Definition 4.2.9. *Given an m-parking function, its m-reading word is obtained by reading from NE to SW line by line, starting from the lines farther from the m-diagonal $x = my$.*

Figure 4.9 illustrates an m-parking function with 132 as its m-reading word. Here $m = 2$. The first line we look at is the line connecting cars 1 and 3. We read

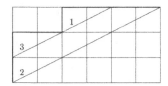

Figure 4.9   A 2-parking function whose 2-reading word is 132.

it from NE to SW so that 1 is before 3. Then the next line to look at is the main diagonal $x = my$ which contains car 2.

Definition 4.2.10. *Given an m-parking function, its natural expansion is defined as follows: starting from (0, 0), each N step, together with the car to its right, is duplicated m times, the car within the N step is duplicated m times and put one to each of the m N steps duplicated; leave each E step untouched.*

Figure 4.10 illustrates the natural expansion of the m-parking function shown in Figure 4.9. Note that the natural expansion of an m-parking function is kind of a "semistandard" parking function in the sense that placing car i immediately on top of car j requires that $i \geq j$ instead of $i > j$.

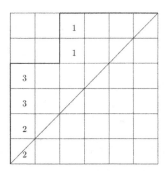

Figure 4.10   The natural expansion of an m-parking function.

Definition 4.2.11. *[Sta99, page 482, Ex. 7.93] For two words $u = (u_1, \ldots, u_k) \in \mathfrak{S}_k$ and $v = (v_1, \ldots, v_l) \in \mathfrak{S}(k+1, k+l)$, where $\mathfrak{S}(m+1, m+l)$ denotes all the permuted words of $\{k+1, \ldots, k+l\}$, $sh(u, v)$ or $sh((u_1, \ldots, u_k), (v_1, \ldots, v_l))$ is the set of shuffles of u and v, i.e., $sh(u, v)$ consists of all permutations $w = (w_1, \ldots, w_{k+l}) \in \mathfrak{S}_{k+l}$ such that both u and v are subsequences of w.*

If the m-reading word of an m-parking function P is a shuffle of the two words $(n - d + 1, \ldots, n)$ and $(n - d, \ldots, 2, 1)$, the increasing order of $(n - d + 1, \ldots, n)$ will imply that any single N segment of P contains at most 1 of $\{n - d + 1, \ldots, n\}$. Furthermore, each of $\{n - d + 1, \ldots, n\}$ should occupy the top spot of some N segment. Hence if we change these d top N steps all to D steps and remove the cars in the m-parking function, we will get an m-Schröder path with d diagonal steps. Conversely, given a path $\Pi \in \mathcal{S}_{n,d}^m$, we may change its d diagonal steps to d NE pairs; after that place cars $\{n - d + 1, \ldots, n\}$ to the right of the d new N steps, and place cars $\{n - d, \ldots, 2, 1\}$ to the right of the other $n - d$ D steps in the uniquely right order so that the m-reading word of the m-parking function formed is a shuffle of the two words $(n - d + 1, \ldots, n)$ and $(n - d, \ldots, 2, 1)$. In this way every m-Schröder path corresponds to an m-parking function of the particular type. Because it is easier to manipulate when there are no D steps, we define the m-Schröder polynomial in the following way.

Definition 4.2.12. *The m-Schröder polynomial is defined as*

$$S_{n,d}^m(q,t) = \sum_{\substack{\Pi:\ \Pi \in \mathbb{P}_n^m \text{ and the } m\text{-reading word of } \Pi \\ \in sh((n-d+1,\ldots,n),(n-d,\ldots,1))}} q^{\text{dinv}_m(\Pi)} t^{\text{area}(\Pi)},$$

where $\text{dinv}_m(\Pi) = \text{dinv}(\hat{\Pi})$, $\hat{\Pi}$ *is the natural expansion of* Π, *and* dinv *is the obvious generalization of the statistic on parking functions introduced in [HL05].*

The following m-Shuffle Theorem may be viewed as a generalization of the original Shuffle Conjecture by Haglund, Haiman, Loehr, Remmel and Ulyanov [HHL+05]. The original Shuffle Conjecture, in terms of combinatorial descriptions of Frobenius character of the diagonal coinvariant algebra R_n in n pairs of variables, was refined in [HMZ12], witnessed significant efforts to prove several special cases of it, and was finally proved by Carlsson and Mellit [CM18].

Theorem 4.2.6. *The "m-Shuffle Theorem". [CM18]*

$$S_{n,d}^m(q,t) = <\nabla^m e_n, e_{n-d} h_d>,$$

where ∇ *is a linear operator defined in terms of the modified Macdonald polynomials.*

For details and proofs of Theorem 4.2.6, see [HHL+05] and [CM18].

Recent updates on combinatorics of rectangular Schröder parking functions are available in [AB18].

4.3 MISCELLANEOUS TOPICS

4.3.1 Lattice Paths Enumerator

Discussed in [SY12], Lattice path enumerator is a way to represent all the lattice paths bounded by fixed boundaries in terms of the area statistic distribution. It turns out to be a useful tool, especially in the investigation of $(\mathrm{des}, \mathrm{inv})$ joint distributions of the lattice paths restricted in a Ferrers board.

If x is a positive integer, a lattice path from the origin $(0,0)$ to the point (x,n) can be encoded by a nondecreasing sequence $(x_1, x_2, \ldots, x_n)$ of length n, where $0 \leq x_i \leq x$ and x_i is the x-coordinate of the ith north step. For example, let $x = 5$ and $n = 3$. Then the path $EENENNEE$ is encoded by $(2,3,3)$.

In general, let s be a nondecreasing sequence with positive integer terms $s_1, s_2, \ldots, s_n$. A lattice path from $(0,0)$ to (x,n) is one with the right boundary s if $x_i < s_i$ for $1 \leq i \leq n$. If $x \geq s_n$, then the number of lattice paths from $(0,0)$ to (x,n) with the right boundary s does not depend on x. Let $Path_n(\mathbf{s})$ be the set of lattice paths from $(0,0)$ to (s_n, n) with the right boundary s, and $LP_n(\mathbf{s})$ be the cardinality of $Path_n(\mathbf{s})$. For a given sequence $\mathbf{s} = (s_1, s_2, \ldots, s_n)$, let the *area enumerator of lattice paths* with right boundary s be defined by,

$$LP_n(\mathbf{s}; q) = \sum_{\Pi \in Path_n(\mathbf{s})} q^{\mathrm{area}(\Pi)},$$

where $\mathrm{area}(\Pi) = \sum_{i=1}^{n} x_i$ is the area enclosed by the path Π, the y-axis, and the line $y = n$. Hence $LP_n(\mathbf{s}) = LP_n(\mathbf{s}; 1)$.

We allow the entries s_i to satisfy $s_1 \geq s_2 \geq \cdots \geq s_n$, say, in which case

$$LP_n(\mathbf{s}; q) = LP_n((s_n, s_n, \ldots, s_n); q) = \begin{bmatrix} s_n + n - 1 \\ n \end{bmatrix}.$$

In particular $LP_n((n+1, n+1, \ldots, n+1); q) = \begin{bmatrix} 2n \\ n \end{bmatrix}$. It is also easy to see that

$$LP_n((1, 2, \ldots, n); q) = C_n(q),$$

where $C_n(q)$ is Carlitz and Riordan's q-Catalan polynomial.

Let F be a Ferrers board with n rows and n columns, which is aligned on the top and left as in Section 3.4. But here for convenience we index the rows from bottom to top, and columns from left to right. Let r_i be the size of row i. Hence $1 \leq r_1 \leq r_2 \leq \cdots \leq r_n = n$.

For a set $D = \{d_1, d_2, \ldots, d_k\}$ with $1 \leq d_1 < \cdots < d_k \leq n - 1$, let $\beta_F(D)$ be the number of permutations in F with the descent set D, and $\alpha_F(D)$ be the number of permutations in F whose descent set is contained in D. The inv q-analogues of $\alpha_F(D)$ and $\beta_F(D)$ are defined by

$$\alpha_F(D,q) = \sum_{\sigma \in F : \mathrm{Des}(\sigma) \subseteq D} q^{\mathrm{inv}(\sigma)}, \qquad \beta_F(D,q) = \sum_{\sigma \in F : \mathrm{Des}(\sigma) = D} q^{\mathrm{inv}(\sigma)}.$$

Clearly $\alpha_F(D,1) = \alpha_F(D)$ and $\beta_F(D,1) = \beta_F(D)$. The Inclusion-Exclusion Principle implies that

$$\alpha_F(D,q) = \sum_{T \subseteq D} \beta_F(T,q), \qquad \beta_F(D,q) = \sum_{T \subseteq D} (-1)^{|D-T|} \alpha_F(T,q).$$

Below we show that $\alpha_F(D,q)$ and $\beta_F(D,q)$ can be expressed in terms of $LP_n(\mathbf{s},q)$, the area enumerator of lattice paths with proper right boundaries and lengths.

Let's first compute $\alpha_F(D,q)$. To get a permutation σ in F satisfying $\mathrm{Des}(\sigma) \subseteq D$, we first choose $x_1 < x_2 < \cdots < x_{d_1}$ such that $1 \leq x_i \leq r_i$, and put a 1 in the cell (x_i, i) for $1 \leq i \leq d_1$. Then choose $x_{d_1+1} < x_{d_1+2} < \cdots < x_{d_2}$ such that $1 \leq x_i \leq r_i$, and put a 1 in the cell (x_i, i) for $d_1 < i \leq d_2$, and so on.

We say that the cell (i,j) is a 1-cell if it is filled with a 1. It is clear that an inversion of σ corresponds to a southeast chain of size 2 in the filling, i.e. a pair of 1-cells $\{ (x_{i_1}, i_1), (x_{i_2}, i_2) \}$ such that $i_1 < i_2$ while $x_{i_1} > x_{i_2}$.

For $1 \leq i \leq d_1$, the 1-cell in the ith row (i.e. $y = i$) has exactly $x_i - i$ many other 1-cells lying above it and to its left. Hence the 1-cell in the ith row contributes $x_i - i$ to the statistic $\mathrm{inv}(\sigma)$, and all the 1-cells in the first d_1 rows contributed

$$(x_1 - 1) + (x_2 - 2) + \cdots + (x_{d_1} - d_1)$$

to the statistic $\mathrm{inv}(\sigma)$.

Note that $0 \leq x_1 - 1 \leq x_2 - 2 \leq \cdots \leq x_{d_1} - d_1$, and $x_i - i < r_i - i + 1$. Hence the number of choices for the sequence $(x_1, \ldots, x_{d_1})$ is exactly the number of lattice paths from $(0,0)$ to $(r_{d_1} - d_1 + 1, d_1)$ with the right boundary $(r_1, r_2 - 1, \ldots, r_{d_1} - d_1 + 1)$, and $\sum_{i=1}^{d_1}(x_i - i)$ is the area of the corresponding lattice path. Therefore the first d_1 rows of F contribute a factor of $LP_{d_1}((h_1, \ldots, h_{d_1}); q)$ to $\alpha_F(D, q)$.

Let $\mathbf{h} = (h_1, h_2, \ldots, h_n)$ where $h_i = r_i - i + 1$. Let the ith block of F consist of rows $d_{i-1} + 1$ to d_i. Applying the above analysis to the ith block of the Ferrers board F for $i = 2, \ldots, k+1$, we get that

Theorem 4.3.1. *For any set $D = \{d_1, d_2, \ldots, d_k\}$ with $1 \le d_1 < \cdots < d_k \le n-1$ and Ferrers board F, let $\alpha_F(D)$ be the number of permutations in F whose descent set is contained in D. Then the* inv *q-analogue of $\alpha_F(D)$ is given by,*

$$\alpha_F(D, q) = \sum_{\sigma \in F: \mathrm{Des}(\sigma) \subseteq D} q^{\mathrm{inv}(\sigma)} = \prod_{i=0}^{k} LP_{d_{i+1} - d_i}((h_{d_i+1}, \ldots, h_{d_{i+1}}); q) \quad (4.6)$$

where we use the convention that $d_0 = 0$ and $d_{k+1} = n$.

In consequence,

$$\beta_F(D, q) = \sum_{T \subseteq D} (-1)^{|D-T|} \alpha_F(T, q)$$

$$= \sum_{1 \le i_1 < i_2 < \cdots < i_j \le k} (-1)^{k-j} f(0, i_1) f(i_1, i_2) \ldots f(i_j, k+1)$$

where

$$f(i, j) = \begin{cases} LP_{d_j - d_i}(h_{d_i+1}, \ldots, h_{d_j}); q) & \text{if } i < j \\ 1 & \text{if } i = j \\ 0 & \text{if } i > j. \end{cases} \quad (4.7)$$

Following the discussion of Stanley [Sta97, p. 69-70], we obtain that

Theorem 4.3.2. *For any set $D = \{d_1, d_2, \ldots, d_k\}$ with $1 \le d_1 < \cdots < d_k \le n-1$ and Ferrers board F, let $\beta_F(D)$ be the number of permutations in F with the descent set D. Then the* inv *q-analogue of $\beta_F(D)$, $\beta_F(D, q) = \sum_{\sigma \in F: \mathrm{Des}(\sigma) = D} q^{\mathrm{inv}(\sigma)}$, is the determinant of a $(k+1) \times (k+1)$ matrix with its (i,j) entry $f(i, j+1)$, $0 \le i, j \le k$. That is,*

$$\beta_F(D, q) = \det[f_{i,j+1}]_0^k \quad (4.8)$$

where $f(i, j)$ is given by (4.7).

When the Ferrers board F is an $n \times n$ square, $h_i = n - i + 1$, and

$$LP_{j-i}((h_{i+1}, \ldots, h_j); q) = \begin{bmatrix} n - i \\ j - i \end{bmatrix}.$$

Theorems 4.3.1 and 4.3.2 reduce to the classical results [Sta97, Example 2.2.5]

$$\sum_{\substack{\pi \in \mathfrak{S}_n \\ Des(\pi) \subseteq D}} q^{\mathrm{inv}(\pi)} = \begin{bmatrix} n \\ d_1, d_2 - d_1, \ldots, n - d_k \end{bmatrix}$$

and

$$\sum_{\substack{\pi \in \mathfrak{S}_n \\ Des(\pi) = D}} q^{\mathrm{inv}(\pi)} = \det \left[\begin{bmatrix} n - d_i \\ d_{j+1} - d_i \end{bmatrix} \right]_0^k.$$

For a general Ferrers board with n rows and n columns, let's check two extreme cases: $D = \{1, 2, \ldots, n - 1\}$ and $D = \emptyset$.

- Case 1. $D = \{1, 2, \ldots, n - 1\}$: Theorem 4.3.1 yields the identity

$$\sum_{\sigma \in F} q^{\mathrm{inv}(\sigma)} = \alpha_F(\{1, 2, \ldots, n - 1\}, q) = \prod_{i=1}^{n} LP_1((h_i); q) = \prod_{i=1}^{n} [h_i]. \quad (4.9)$$

Note that permutation fillings of a Ferrers board with n rows and n columns correspond to complete matchings of $\{1, \ldots, 2n\}$ with fixed sets of left endpoints and right endpoints, and an inversion of the permutation is exactly a nesting of the matching; See de Mier [dM06] and Kasraoui [Kas10]. To see this, for a given Ferrers board F with n rows and n columns, one traverses the path from the lower-left corner to the top-right corner, and records the path by its steps $a_1, a_2, \ldots, a_{2n}$ where $a_i = E$ if the ith step is East, and $a_i = N$ if the ith step is North. Let $L = \{i : a_i = E\}$ and $R = \{i : a_i = N\}$. Then 01-fillings of F considered here are in one-to-one correspondence with the matchings of $\{1, \ldots, 2n\}$ for which the set of left endpoints is L and the set of right endpoints is R. For example, in the following Ferrers board F, traversing from the lower-left corner to the top-right corner, we get the sequence $EENENENN$. Thus $L = 1, 2, 4, 6$ and $R = 3, 5, 7, 8$. The filling given in the figure corresponds to the matching $\{(1, 7), (2, 3), (4, 5), (6, 8)\}$.

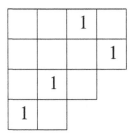

Figure 4.11 (Truncated) board representation of the permutation 1243.

It follows that equation (4.9) is exactly the generating function of the statistic $ne_2(M)$, which is the number of nestings in a matching M, counted over all the matchings with given sets of left and right endpoints. That is,

$$\sum_M q^{ne_2(M)} = \prod_{i=1}^n [h_i],$$

which matches the known results in [dSC93, Kas10].

- Case 2, $D = \emptyset$: We have

$$\alpha_F(\emptyset, q) = \beta_F(\emptyset, q) = LP_n((h_1, \ldots, h_n), q)$$

Note that $h_n = 1$, hence $LP_n((h_1, \ldots, h_n), q) = 1$ iff $r_i \geq i$ for all i, where the only permutation in the Ferrers board F with no descents is the identity permutation; otherwise $Path_n(h_1, \ldots, h_n) = \emptyset$ and $LP_n((h_1, \ldots, h_n), q) = 0$.

Theorems 4.3.1 and 4.3.2 may be used to get a formula for the joint distribution of $\mathrm{des}(\sigma)$ and $\mathrm{inv}(\sigma)$ over permutations in F. Let

$$A_F(t, q) = \sum_{\sigma \in F} t^{1+\mathrm{des}(\sigma)} q^{\mathrm{inv}(\sigma)}. \tag{4.10}$$

Theorem 4.3.3.
$$A_F(t, q) = (1 - t)^n per(M), \tag{4.11}$$

where M is an $n \times n$ matrix whose (i, j)-entry is given by

$$M_{ij} = \begin{cases} \frac{t}{1-t} LP_{j-i+1}((h_i, \ldots, h_j); q) & \text{if } i \leq j \\ 1 & \text{if } i = j+1 \\ 0 & \text{if } i > j+1, \end{cases}$$

and $per(M)$ is the permanent of the matrix M.

Proof. We have

$$
\begin{aligned}
\sum_{\sigma \in F} t^{1+\mathrm{des}(\sigma)} q^{\mathrm{inv}(\sigma)} &= \sum_{D \subseteq \{1,2,\ldots,n-1\}} t^{1+|D|} \beta_F(D,q) \\
&= \sum_{D \subseteq \{1,2,\ldots,n-1\}} t^{1+|D|} \sum_{T:T \subseteq D} (-1)^{|D-T|} \alpha_F(T,q) \\
&= \sum_{T \subseteq \{1,2,\ldots,n-1\}} \alpha_F(T,q) \sum_{D:T \subseteq D} (-1)^{|D-T|} t^{1+|T|+|D-T|} \\
&= (1-t)^n \sum_{T=\{t_1,\ldots,t_k\}_<} \left(\frac{t}{1-t}\right)^{1+k} \Delta_T(LP_n(\mathbf{h})) \\
&= (1-t)^n \mathrm{per}(M),
\end{aligned}
$$

where M is the $n \times n$ matrix as described in Theorem 4.3.3, and $\Delta_D(LP_n(\mathbf{h}))$ denotes the right-hand side of (4.6). □

Remark 4.3.1. The statistic of des of permutations in a Ferrers board provides an example of one-dependent determinantal point processes, as studied by Borodin, Diaconis, and Fulman [BDF10]. Let U be a finite set. A point process on U is a probability measure P on the $2^{|U|}$ subsets of U. One simple way to specify P is via its correlation functions $\rho(A)$, where for $A \subseteq U$,

$$
\rho(A) = P\{S : S \supseteq A\}.
$$

A point process is *determinantal* with kernel $K(x,y)$ if

$$
\rho(A) = \det[K(x,y)]_{x,y \in A}.
$$

It is *one-dependent* if $\rho(X \cup Y) = \rho(X)\rho(Y)$, whenever dist $(X,Y) \geq 2$.

Borodin et al. showed that many examples from combinatorics, algebra, and group theory are determinantal one-dependent point processes, for example, the carries process, the descent set of uniformly random permutations, and the descent set in Mallows model [BDF10]. For these three cases, the point processes are stationary, while the descent set of permutations in a Ferrers board corresponds to a determinantal one-dependent point process that is not stationary. Explicitly, for any set $D = \{d_1, \ldots, d_k\}$ with $1 \leq d_1 < \cdots < d_k \leq n-1$, let $P_F(D) = \beta_F(D)/(\prod_{i=1}^{n} h_i)$. Using [BDF10, Theorem 7.5] we obtain that P_F is a *determinantal, one-dependent process with correlation functions*

$$\rho(D) = \alpha_F(D) = \det[K(d_i, d_j)]_{i,j=1}^k$$

and with correlation kernel

$$K(x, y) = \delta_{x,y} + (E^{-1})_{x,y+1},$$

where E is the upper triangular matrix $E = [e(i-1,j)]_{i,j=1}^n$, whose entries are given by

$$e(i, j) = \begin{cases} LP_{j-i}(h_{i+1}, \ldots, h_j) & \text{if } i < j \\ 1 & \text{if } i = j \\ 0 & \text{if } i > j. \end{cases}$$

4.3.2 Lattice Paths of General Steps

In this subsection we introduce lattice paths of more general steps. In fact, lattice paths may be generalized in several directions: the step set, domain of the paths, and dimension of the space. In [HZZ16] (see also [Zha15]), a set of general paths called $S - (p, q)$ paths is studied.

Let S be a multiset in $\mathbb{N}^2 \setminus \{(0,0)\}$. An S-(p,q)-*path* is a lattice path from $(0,0)$ to (pn, qn) which never goes below the line $py = qx$ with step set S, where $p, q, n \in \mathbb{N}$, $\gcd(p,q) = 1$,

Thus $\{(0,1), (1,0)\}$-$(1,1)$-paths are the classical Dyck paths, with $S = \{(0,1), (1,0)\}$ and $(p,q) = (1,1)$. Other examples include the $\{(0,1), (1,0)\}$-$(1,k)$-paths [HP91], $\{(0,1), (1,0)\}$-(p,q)-paths [Sat89], $\{(k,k), (0,2), (2,0)\}$-$(1,1)$-paths by some rotations [MS08], S-$(1,1)$-paths for certain special step sets S [Ruk11, KdM13], and general S-(p,q)-paths of which solutions are given by generating functions satisfying given equations [LY90, Duc00].

In [HZZ16], there is a general treatment of the S-(p,q)-paths: fix the step set $S = \{\mathbf{u_i} = (a_i, b_i) \mid i \in \gamma\} \subseteq \mathbb{N}^2 \setminus \{(0,0)\}$, where p and q are fixed co-prime positive integers and all variables are nonnegative.

Notation 4.3.2. *Among the paths from $(0,0)$ to (pn, qn), let $\mathcal{C}_{pn,qn}^S$ denote the set of all S-(p,q) paths, let $\mathcal{B}_{pn,qn}^S$ denote the set of all S-(p,q) paths meeting the diagonal only at the two ends, and $\mathcal{C}_{pn,qn,t}^S$ denote the set of all S-(p,q) paths meeting with* *the diagonal exactly t times beyond* $(0,0)$. *Let* $C^S_{pn,qn}$, $B^S_{pn,qn}$ *and* $C^S_{pn,qn,t}$ *denote* $|\mathcal{C}^S_{pn,qn}|$, $|\mathcal{B}^S_{pn,qn}|$ *and* $|\mathcal{C}^S_{pn,qn,t}|$, *respectively.*

Definition 4.3.3. *For fixed S, define the **class** of a path π to be a sequence $\alpha = (\alpha_i)_{i \in \gamma}$, denoted by $\alpha(\pi)$, where α_i is the number of $\mathbf{u_i}$-steps in π (without loss of generality we regard the elements in S as ordered alphabetically, thus α is a sequence).*

Class sequence describes the distribution of steps in π. All the $\{(0,1), (1,0)\}$-(p,q)-paths from $(0,0)$ to (pn, qn) have a unique class (qn, pn), but there may be various classes for general S.

Notation 4.3.4. *For $\alpha \in \mathbb{N}^\gamma$, let $C^S_\alpha, C^S_{\alpha,t}$ and B^S_α denote the number of paths in $\bigcup_{n=0}^\infty \mathcal{C}^S_{pn,qn}$, $\bigcup_{n=0}^\infty \mathcal{C}^S_{pn,qn,t}$ and $\bigcup_{n=0}^\infty \mathcal{B}^S_{pn,qn}$ with class α, respectively.*

Notation 4.3.5. *Let*

$$\mathcal{A}^n = \mathcal{A}^n_{p,q} := \{(\alpha_i)_{i \in \gamma} \in \mathbb{N}^\gamma \mid \sum_{i \in \gamma} \alpha_i(a_i, b_i) = (pn, qn)\}$$

and $\mathcal{A} = \mathcal{A}_{p,q} := \bigcup_{n=0}^\infty \mathcal{A}^n_{p,q}$.

Notation 4.3.6. *For a multiset M in $\mathbb{N}$ satisfying $\sum_{m \in M} m < \infty$, let*

$$a_M = \frac{1}{\sum_{m \in M} m} \frac{(\sum_{m \in M} m)!}{\prod_{m \in M} m!}$$

and $a_{pn,qn} = \sum_{\alpha \in \mathcal{A}^n} a_\alpha$. *(Define $a_M = 0$ when $\sum_{m \in M} m = 0$.)*

Theorem 4.3.4. *[HZZ16] For $n \geq t > 0$, $\alpha \in \mathcal{A}^n_{p,q}$, there holds:*

(i) $C(x) = e^{A(x)} - 1$. *Thus* $C_{pn,qn} = \sum_{n_1 + 2n_2 + 3n_3 + \cdots = n} \prod_{i=1}^\infty \frac{a_{pi,qi}^{n_i}}{n_i!}$.

(ii) $B(x) = 1 - e^{-A(x)}$. *Thus* $B_{pn,qn} = \sum_{n_1 + 2n_2 + 3n_3 + \cdots = n} (-1)^{1 + \sum_{i=1}^\infty n_i} \prod_{i=1}^\infty \frac{a_{pi,qi}^{n_i}}{n_i!}$.

(iii) $C(x,u) = e^{A(x,u)} - 1$. *Thus* $C_\alpha = \sum_{\sum_{\lambda \in \mathcal{A}} \lambda n_\lambda = \alpha} \prod_{\lambda \in \mathcal{A}} \frac{a_\lambda^{n_\lambda}}{n_\lambda!}$.

(iv) $B(x,u) = 1 - e^{-A(x,u)}$. *Thus* $B_\alpha = \sum_{\sum_{\lambda \in \mathcal{A}} \lambda n_\lambda = \alpha} (-1)^{1 + \sum_{\lambda \in \mathcal{A}} n_\lambda} \prod_{\lambda \in \mathcal{A}} \frac{a_\lambda^{n_\lambda}}{n_\lambda!}$.

The proof of Theorem 4.3.4 is via careful analysis. Although Theorem 4.3.4 appears to be complicated, it has appealing applications for enumerating lattice paths.

Theorem 4.3.5. *For $\alpha \in \mathcal{A}^1_{p,q}$, we have:*

(i) *The number of (p, q)-paths from $(0, 0)$ to (p, q) is*

$$C_{p,q} = B_{p,q} = a_{p,q} = \sum_{\alpha \in \mathcal{A}^1} \frac{1}{\sum_{i \in \gamma} \alpha_i} \frac{(\sum_{i \in \gamma} \alpha_i)!}{\prod_{i \in \gamma} \alpha_i!}.$$

(ii) *The number of (p, q)-paths from $(0, 0)$ to (p, q) with class α is*

$$C_{\alpha} = B_{\alpha} = a_{\alpha} = \frac{1}{\sum_{i \in \gamma} \alpha_i} \frac{(\sum_{i \in \gamma} \alpha_i)!}{\prod_{i \in \gamma} \alpha_i!}.$$

The following lemma deduced from Theorem 4.3.5 shows that $(m, 1)$-paths from $(0, 0)$ to (mn, n) has a close relation with $(mn + 1, n)$-paths:

Lemma 4.3.7. *Suppose $(1, 0) \in S$, $m, n > 0$. Besides, for any $\mathbf{u_i} = (a_i, b_i) \in S \setminus \{(1, 0)\}$, we have $mb_i \geq a_i$. Then $C_{mn,n} = a_{mn+1,n}$ and $C_{\alpha'} = a_\alpha$, where $\alpha \in \mathcal{A}^1_{mn+1,n}$,*

$$\alpha'_i = \begin{cases} \alpha_i - 1, & \text{if } \mathbf{u_i} = (1, 0), \\ \alpha_i, & \text{otherwise.} \end{cases}$$

Proof. $\forall\, 0 < i < n$, the segment $\{(x, i) \mid mi < x \leq mi + \frac{i}{n}, x \in \mathbb{R}\}$ contains no integral points. Hence, there are no integral points in the segment $\{(x, \frac{nx}{mn+1}) \mid 0 < x < mn + 1, x \in \mathbb{R}\}$ and the interior of the area surround by $y = \frac{x}{m}, y = \frac{nx}{mn+1}$ and $y = n$. By the geometric meaning of the conditions, any $(mn+1, n)$-path from $(0, 0)$ to $(mn + 1, n)$ is composed of an $(m, 1)$-path from $(0, 0)$ to (mn, n) plus a $(1, 0)$-step. Since $\gcd(mn + 1, n) = 1$, we get the conclusion immediately from Theorem 4.3.5. □

Example 4.3.8. *Let $S = \{(0, 1), (1, 0), (1, 1)\}$. Then*

$$\mathcal{A}^n = \{(qn - i, pn - i, i) \mid i \leq \min\{pn, qn\}, i \in \mathbb{N}\}.$$

Applying Theorem 4.3.5, the number of (p, q)-paths from $(0, 0)$ to (p, q) with class $(q - i, p - i, i)$ (i.e., with i diagonal steps) is

$$C_{(q-i,p-i,i)} = a_{(q-i,p-i,i)} = \frac{1}{p + q - i} \binom{p + q - i}{p - i, q - i, i},$$

and the number of (p, q)-paths from $(0, 0)$ to (p, q) is

$$C_{p,q} = a_{p,q} = \sum_{i=0}^{q} \frac{1}{p + q - i} \binom{p + q - i}{p - i, q - i, i}.$$

Now let $(p, q) = (mn + 1, n)$. Applying Lemma 4.3.7 we have that the number of $(m, 1)$-paths from $(0, 0)$ to (mn, n) with exactly d diagonal steps is equal to

$$C_{(n-i,mn-i,i)} = a_{(n-i,mn+1-i,i)} = \frac{1}{mn - i + 1} \binom{mn + n - i}{mn - i, n - i, i},$$

and the number of all $(m, 1)$-paths from $(0, 0)$ to (mn, n) is

$$C_{mn,n} = \sum_{i=0}^{n} \frac{1}{mn - i + 1} \binom{mn + n - i}{mn - i, n - i, i}.$$

The number $C_{mn,n}$ is indeed the m-Schröder number S_n^m introduced in [Son05a].

4.3.3 Truncated Rectangle

Another direction of lattice paths studies is to extend the domain where interested lattice paths reside. Here we shall investigate the truncated rectangle.

Definition 4.3.9. *Give nonnegative integers n, i, j. A truncated rectangle, denoted by $R(n, i, j)$, includes the boundary and interior part of the region surrounded by the lines: $x = 0, y = 0, y = x - i, x = n + i, y = n + j$, In other words, $R(n, i, j)$ denotes the rectangle decided by the corners set $\{(0, 0), (n + i, 0), (0, n + j), (n + i, n + j)\}$ but with an n by n right triangle cut off from its southeast corner.*

Below we fix the step set $S = \{(0, 1), (1, 0)\}$. A lattice path in the truncated rectangle $R(n, i, j)$ is one from $(0, 0)$ to $(n + i, n + j)$ that resides entirely in $R(n, i, j)$ with steps $(0, 1)$ and $(1, 0)$, that never goes below the line $y = x - i$. The collection of all lattice paths in the truncated rectangle $R(n, i, j)$ is denoted by $\Pi(n, i, j)$. Clearly, $\Pi(n, 0, 0)$ is the set of n-Dyck paths $\mathcal{D}_n$.

Example 4.3.10. *For* $n = 1, i = 1, j = 1$, *there are five lattice paths in the truncated rectangle* $R(1, 1, 1)$.

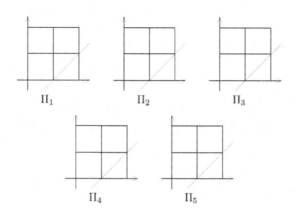

Figure 4.12   The five paths of $\Pi(1, 1, 1)$.

Like before, every path $\Pi \in \Pi(n, i, j)$ has a natural one-to-one correspondence to a $\{0, 1\}$ word $w = w(\Pi)$, by replacing each north step by 0 and each east step by 1. Thus a word corresponding to a path $\Pi \in \Pi(n, i, j)$ has length $l = 2n + i + j$, consisting of $n + j$ 0's and $n + i$ 1's.

The following theorem is a generalization of (4.5). Note that (4.12) reduces to the desired form when i and j are both set to 0.

Theorem 4.3.6. *[HS19] For any nonnegative integers* n, i *and* j, *we have*

$$\sum_{\Pi \in \Pi(n,i,j):\ \mathrm{des}(\Pi)=k} q^{\mathrm{maj}(\Pi)} = q^{k^2} \begin{bmatrix} n+i \\ k \end{bmatrix} \begin{bmatrix} n+j \\ k \end{bmatrix} - q^{i+k^2} \begin{bmatrix} n+1 \\ k+1 \end{bmatrix} \begin{bmatrix} n+i+j-1 \\ k-1 \end{bmatrix}.$$

(4.12)

The proof of Theorem 4.3.6 makes use of Lemma 2.1.1 and (2.2) of Lemma 2.1.2 introduced in Chapter 2, and for details please refer to [HS19].

4.3.4   Multiple Paths

In this subsection we introduce in brief multiple lattice paths which are noncrossing or are nonintersecting.

Definition 4.3.11. *A pair* (Π_1, Π_2) *of lattice paths is* noncrossing *if they have the same origin and the same destination, and* Π_1 *never goes below* Π_2. *In general, lattice paths* $(\Pi_1, \Pi_2, \ldots, \Pi_k)$ *are* noncrossing *if every pair of them is noncrossing.*

The first celebrated result about noncrossing path is the Gessel-Viennot determinant.

Theorem 4.3.7. *[GV85] Let* $0 \le a_1 \le \cdots \le a_k$ *and* $0 \le b_1 \le \cdots \le b_k$ *be two strictly increasing sequences of nonnegative integers. Consider the lattice points* $A_i = (0, a_i)$ *and* $B_i = (b_i, b_i)$, $i \in [k]$. *Then the number of* k-*tuples of noncrossing lattice paths* $(\Pi_1, \Pi_2, \ldots, \Pi_k)$ *such that each* P_i *goes from* A_i *to* B_i *with only* S *or* E *steps is counted by the determinant*

$$\det \begin{pmatrix} a_1, \ldots, a_k \\ b_1, \ldots, b_k \end{pmatrix},$$

where $\begin{pmatrix} a_1, \ldots, a_k \\ b_1, \ldots, b_k \end{pmatrix}$ *represents the* k *by* k *matrix with its* (i, j) *entry* $\binom{a_i}{b_j}$ *(for this reason the determinant is called a "binomial determinant").*

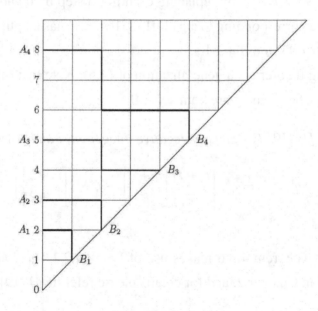

Figure 4.13   The binomial determinant $\begin{pmatrix} 2,3,5,8 \\ 1,2,4,5 \end{pmatrix}$.

Remark 4.3.12. *Note that in the setting of Theorem 4.3.7, the steps are not the usual* $\{N, E\}$. *In fact, the result will be different by using the usual steps set.*

As stuff in this section is cutting-edge and the proofs are complicated, the readers may either be happy with the conclusions only or check the references themselves. We will skip the proofs.

In [Lab93], the author Labelle converts a pair of noncrossing generalized Dyck paths into a single Dyck path and then uses his previous results collaborated with Yeh [LY90] to count these paths. One of the corollaries of [CDD+07] is that the number of k-noncrossing partitions of $\{1, \ldots, n\}$ equals the number of k-nonnesting partitions of $\{1, \ldots, n\}$. Using a bijection between partitions and vacillating tableaux, in [CPQS09], correspondence between pairs of noncrossing free Dyck paths of length $2n$ and noncrossing partitions of $\{1, \ldots, 2n+1\}$ with exactly $n+1$ blocks is established. Below we list a few more interesting results without proofs.

Definition 4.3.13. *Let $k \geq 2$ be an integer. A k-crossing of a matching* M *is a set of k arcs* $(i_{r_1}, j_{r_1}), (i_{r_2}, j_{r_2}), \ldots, (i_{r_k}, j_{r_k})$ *of* M *such that* $i_{r_1} < i_{r_2} < \cdots < i_{r_k} < j_{r_1} < j_{r_2} < \cdots < j_{r_k}$. *A matching without any k-crossing is a k-noncrossing matching.*

Let $f_k(n)$ be the number of k-noncrossing matchings of $\{1, \ldots, 2n\}$ and set

$$F_k(x) = \sum_{n \geq 0} f_k(n) \frac{x^{2n}}{(2n)!},$$

with $f_k(0) = 1$.

Theorem 4.3.8. *[CDD+07]*

1. *The set of 2-noncrossing matchings of $\{1, \ldots, 2n\}$ is in one-to-one correspondence with $\mathcal{D}_n$, the set of n-Dyck paths.*

2. *The set of 3-noncrossing matchings of $\{1, \ldots, 2n\}$ is in one-to-one correspondence with the set of pairs of noncrossing paths in $\mathcal{D}_n$.*

Corollary 4.3.14.

$$f_2(n) = C_n,$$
$$f_3(n) = C_n C_{n+2} - C_{n+1}^2, \tag{4.13}$$

where (4.13) is a result obtained by Gouyou-Beauchamps [GB89].

The proof of Theorem 4.3.8 is based on a bijection between vacillating tableaux and noncrossing partitions. As a matter of fact, a deterministic expression for $f_k(n)$ has been given in [GM93] (see also [CDD$^+$07] for detailed explanations). Furthermore, in [CPQS09], a one-to-one correspondence is established between the set of pairs of noncrossing "free Dyck paths" of length $2n$ (i.e. central Delannoy paths of order n and with 0 diagonal steps) and the set of noncrossing partitions of the set $\{1, \ldots, 2n+1\}$ with $n+1$ blocks.

Note that noncrossing paths are allowed to intersect at turning points, and even to overlap for steps. The extreme situation is that they are just the same paths. Instead, the following definition requires the paths to be both noncrossing and *separate*.

Definition 4.3.15. *In general, lattice paths* $(\Pi_1, \Pi_2, \ldots, \Pi_k)$ *are* nonintersecting *if they have no points in common.*

In Definition 4.3.15, the requirement "the same origin and the same destination" is removed so that the lattice paths are required to not intersect even at the initial and ending points. The difference is nonessential, though. Now for nonnegative integers a, b, c, i and j satisfying $a \le c$ and $d \le b$, let $N(a, b, c, d, i, j)$ denote the set of pairs of nonintersecting lattice paths, (Π_1, Π_2), where each left path Π_1 runs from $(0, 1)$ to $(a, b+1)$ and has i consecutive NE pairs (i.e., left-hand turns), and where each right path Π_2 runs from $(1, 0)$ to $(c+1, d)$ and has j consecutive EN pairs (i.e., right-hand turns). Like before, $\mathrm{maj}(\Pi = \pi_1 \pi_2 \cdots \pi_n) := \sum_k k \cdot \delta(\pi_k > \pi_{k+1})$. Similarly, we also define $\mathrm{less}(\Pi = \pi_1 \pi_2 \cdots \pi_n) := \sum_k k \cdot \delta(\pi_k < \pi_{k+1})$. (Here we adopt the commonly used notation: $\delta(P) = 1$ if the statement P is true and $\delta(P) = 0$ if P is false.) The following theorem due to Krattenthaler and Sulanke, proved by inductive bijection, is a q-analogue of a variant of the distribution formula of Kreweras and Poupard (see [KP86] and [Kre86]).

Theorem 4.3.9.

$$\sum_{(\Pi_1, \Pi_2) \in N(a,b,c,d,i,j)} q^{\mathrm{maj}(\Pi_1)+\mathrm{less}(\Pi_2)}$$
$$= q^{i^2+j^2} \left(\begin{bmatrix} a \\ i \end{bmatrix} \begin{bmatrix} b \\ i \end{bmatrix} \begin{bmatrix} c \\ j \end{bmatrix} \begin{bmatrix} d \\ j \end{bmatrix} - \begin{bmatrix} a+1 \\ i+1 \end{bmatrix} \begin{bmatrix} b-1 \\ i-1 \end{bmatrix} \begin{bmatrix} c-1 \\ j-1 \end{bmatrix} \begin{bmatrix} d+1 \\ j+1 \end{bmatrix} \right).$$

For proof of Theorem 4.3.9, refer to [KS96].

Bibliography

[AB03]   Scott Ahlgren and Matthew Boylan. Arithmetic properties of the par-
           tition function. *Invent. Math.*, 153(3):487–502, 2003.

[AB18]   Jean-Christophe Aval and Francois Bergeron. A note on: rectangu-
           lar Schröder parking functions combinatorics. *Sém. Lothar. Combin.*,
           79:Art. B79a, 13, 2018.

[Aig01]   Martin Aigner. Lattice paths and determinants. In *Computational
           discrete mathematics*, volume 2122 of *Lecture Notes in Comput. Sci.*,
           pages 1–12. Springer, Berlin, 2001.

[AK19a]   George E. Andrews and Christian Krattenthaler, editors. *Lattice path
           combinatorics and applications*, volume 58 of *Developments in Math-
           ematics*. Springer, Cham, 2019.

[AK19b]   Shaun Ault and Charles Kicey. *Counting lattice paths using
           Fourier methods*. Applied and Numerical Harmonic Analysis.
           Birkhäuser/Springer, Cham, 2019. Lecture Notes in Applied and Nu-
           merical Harmonic Analysis.

[Alo02]   Noga Alon. Discrete mathematics: methods and challenges. In *Pro-
           ceedings of the International Congress of Mathematicians, Vol. I (Bei-
           jing, 2002)*, pages 119–135. Higher Ed. Press, Beijing, 2002.

[And87]   D. André. Solution directe du problème résolu par M. Bertrand.
           Comptes Rendus Acad. Sci. Paris, 105:436–437, 1887.

[And98]   George E. Andrews. *The theory of partitions*. Cambridge Mathematical
           Library. Cambridge University Press, Cambridge, 1998. Reprint of the
           1976 original.

[Ava08]   Jean-Christophe Aval. Multivariate Fuss-Catalan numbers. *Discrete Math.*, 308(20):4660–4669, 2008.

[AZ18]   Martin Aigner and Günter M. Ziegler. *Proofs from The Book.* Springer, Berlin, sixth edition, 2018. See corrected reprint of the 1998 original [MR1723092], Including illustrations by Karl H. Hofmann.

[Bí5]   Miklós Bóna, editor. *Handbook of enumerative combinatorics.* Discrete Mathematics and its Applications (Boca Raton). CRC Press, Boca Raton, FL, 2015.

[BDF10]   Alexei Borodin, Persi Diaconis, and Jason Fulman. On adding a list of numbers (and other one-dependent determinantal processes). *Bull. Amer. Math. Soc. (N.S.)*, 47(4):639–670, 2010.

[Ber72]   J. Bertrand. *Calcul des probabilités.* Chelsea Publishing Co., Bronx, N.Y., 1972. Réimpression de la deuxième édition de 1907.

[BK01]   Jason Bandlow and Kendra Killpatrick. An area-to-inv bijection between Dyck paths and 312-avoiding permutations. *Electron. J. Combin.*, 8(1):Research Paper 40, 16 pp. (electronic), 2001.

[BKK⁺19]   Cyril Banderier, Christian Krattenthaler, Alan Krinik, Dmitry Kruchinin, Vladimir Kruchinin, David Nguyen, and Michael Wallner. Explicit formulas for enumeration of lattice paths: basketball and the kernel method. In *Lattice path combinatorics and applications*, volume 58 of *Dev. Math.*, pages 78–118. Springer, Cham, 2019.

[BKR17]   A. Bostan, I. Kurkova, and K. Raschel. A human proof of Gessel's lattice path conjecture. *Trans. Amer. Math. Soc.*, 369(2):1365–1393, 2017.

[BMM10]   Mireille Bousquet-Mélou and Marni Mishna. Walks with small steps in the quarter plane. In *Algorithmic probability and combinatorics*, volume 520 of *Contemp. Math.*, pages 1–39. Amer. Math. Soc., Providence, RI, 2010.

[BQ03]   Arthur T. Benjamin and Jennifer J. Quinn. *Proofs that really count*, volume 27 of *The Dolciani Mathematical Expositions*. Mathematical Association of America, Washington, DC, 2003. The art of combinatorial proof.

[Bru04]   Richard A. Brualdi. *Introductory combinatorics*. Prentice Hall, Upper Saddle River, NJ, 4th edition, 2004.

[BSS93]   Joseph Bonin, Louis Shapiro, and Rodica Simion. Some q-analogues of the Schröder numbers arising from combinatorial statistics on lattice paths. *J. Statist. Plann. Inference*, 34(1):35–55, 1993.

[BWX07]   Jörgen Backelin, Julian West, and Guoce Xin. Wilf-equivalence for singleton classes. *Adv. in Appl. Math.*, 38(2):133–148, 2007.

[BZ85]   David M. Bressoud and Doron Zeilberger. Bijecting Euler's partitions-recurrence. *Amer. Math. Monthly*, 92(1):54–55, 1985.

[Cam]   Peter J. Cameron. Combinatorics entering the third millennium. Available at www.maths.qmul.ac.uk/~pjc/preprints/pfhist.pdf.

[Car54]   L. Carlitz. q-Bernoulli and Eulerian numbers. *Trans. Amer. Math. Soc.*, 76:332–350, 1954.

[CDD+07]   William Y. C. Chen, Eva Y. P. Deng, Rosena R. X. Du, Richard P. Stanley, and Catherine H. Yan. Crossings and nestings of matchings and partitions. *Trans. Amer. Math. Soc.*, 359(4):1555–1575, 2007.

[Charalam] Charalambos A. Charalambides. *Enumerative combinatorics*. CRC Press Series on Discrete Mathematics and Its Applications. Chapman & Hall/CRC, Boca Raton, FL, 2002.

[Cig87]   J. Cigler. Some remarks on Catalan families. *European J. Combin.*, 8(3):261–267, 1987.

[CM18]   Erik Carlsson and Anton Mellit. A proof of the shuffle conjecture. *J. Amer. Math. Soc.*, 31(3):661–697, 2018.

[Com74]   Louis Comtet. *Advanced combinatorics*. D. Reidel Publishing Co., Dordrecht, enlarged edition, 1974. The art of finite and infinite expansions.

[Coo49]   J. L. Coolidge. The story of the binomial theorem. *Amer. Math. Monthly*, 56:147–157, 1949.

[CPQS09]   William Y. C. Chen, Sabrina X. M. Pang, Ellen X. Y. Qu, and Richard P. Stanley. Pairs of noncrossing free Dyck paths and noncrossing partitions. *Discrete Math.*, 309(9):2834–2838, 2009.

[CR64]   L. Carlitz and J. Riordan. Two element lattice permutation numbers and their q-generalization. *Duke Math. J.*, 31:371–388, 1964.

[CU19]   Mahir Bilen Can and Özlem Uğurlu. The genesis of involutions (polarizations and lattice paths). *Discrete Math.*, 342(1):201–216, 2019.

[CW]   Xin Chen and Jane Wang. The super catalan numbers $s(m, m + s)$ for $s \leq 4$. *arXiv preprint arXiv:1208.4196*, 2012.

[Del95]   H. Delannoy. Emploi de l'échiquier pour la résolution de certains problèmes de probabilités. *Assoc. Franc. Bordeaux*, 24:70–90, 1895.

[DFZ19]   Rosena R. X. Du, Xiaojie Fan, and Yue Zhao. Enumeration on row-increasing tableaux of shape $2 \times n$. *Electron. J. Combin.*, 26(1):Paper 1.48, 13, 2019.

[Dil08]   Karl Dilcher. Determinant expressions for q-harmonic congruences and degenerate Bernoulli numbers. *Electron. J. Combin.*, 15(1):Research paper 63, 18, 2008.

[dM06]   Anna de Mier. On the symmetry of the distribution of k-crossings and k-nestings in graphs. *Electron. J. Combin.*, 13(1):Note 21, 6 pp. (electronic), 2006.

[dSC93]   M. de Sainte-Catherine. Couplages et pfaffiens en combinatoire, physique et informatique. Master's thesis, University of Bordeaux I, Talence, France, 1993.

[Duc00]   Philippe Duchon. On the enumeration and generation of generalized Dyck words. *Discrete Math.*, 225(1-3):121–135, 2000. Formal power series and algebraic combinatorics (Toronto, ON, 1998).

[EF08]   Sen-Peng Eu and Tung-Shan Fu. Lattice paths and generalized cluster complexes. *J. Combin. Theory Ser. A*, 115(7):1183–1210, 2008.

[EHKK03] E. S. Egge, J. Haglund, K. Killpatrick, and D. Kremer. A Schröder generalization of Haglund's statistic on Catalan paths. *Electron. J. Combin.*, 10(1):Research Paper 16, 21 pp. (electronic), 2003.

[EL91]   Arulappah Eswarathasan and Eugene Levine. p-integral harmonic sums. *Discrete Math.*, 91(3):249–257, 1991.

[Eul45]   Leonhard Euler. *Introductio in Analysin Infinitorum. (Opera Omnia. Series Prima: Opera Mathematica, Volumen Novum.)*. Societas Scientiarum Naturalium Helveticae, Geneva, 1945. Editit Andreas Speiser.

[FH85]   J. Fürlinger and J. Hofbauer. q-Catalan numbers. *J. Combin. Theory Ser. A*, 40(2):248–264, 1985.

[FRT54]   J. S. Frame, G. de B. Robinson, and R. M. Thrall. The hook graphs of the symmetric groups. *Canad. J. Math.*, 6:316–324, 1954.

[FS78]   Dominique Foata and Marcel-Paul Schützenberger. Major index and inversion number of permutations. *Math. Nachr.*, 83:143–159, 1978.

[FSV06]   Philippe Flajolet, Wojciech Szpankowski, and Brigitte Vallée. Hidden word statistics. *J. ACM*, 53(1):147–183, 2006.

[Ful97]   William Fulton. *Young tableaux*, volume 35 of *London Mathematical Society Student Texts*. Cambridge University Press, Cambridge, 1997. With applications to representation theory and geometry.

[GB89]   Dominique Gouyou-Beauchamps. Standard Young tableaux of height 4 and 5. *European J. Combin.*, 10(1):69–82, 1989.

[GBGL08] Timothy Gowers, June Barrow-Green, and Imre Leader, editors. *The Princeton companion to mathematics*. Princeton University Press, Princeton, NJ, 2008.

[Ges80] Ira M. Gessel. A factorization for formal Laurent series and lattice path enumeration. *J. Combin. Theory Ser. A*, 28(3):321–337, 1980.

[Ges86] Ira M. Gessel. A probabilistic method for lattice path enumeration. *J. Statist. Plann. Inference*, 14(1):49–58, 1986.

[Ges92] Ira M. Gessel. Super ballot numbers. *J. Symbolic Comput.*, 14(2-3):179–194, 1992.

[GG79] A. M. Garsia and I. Gessel. Permutation statistics and partitions. *Adv. in Math.*, 31(3):288–305, 1979.

[GH96] A. M. Garsia and M. Haiman. A remarkable q,t-Catalan sequence and q-Lagrange inversion. *J. Algebraic Combin.*, 5(3):191–244, 1996.

[GH01] A. M. Garsia and J. Haglund. A positivity result in the theory of Macdonald polynomials. *Proc. Natl. Acad. Sci. USA*, 98(8):4313–4316 (electronic), 2001.

[GH02] A. M. Garsia and J. Haglund. A proof of the q,t-Catalan positivity conjecture. *Discrete Math.*, 256(3):677–717, 2002. LaCIM 2000 Conference on Combinatorics, Computer Science and Applications (Montreal, QC).

[GKP89] Ronald L. Graham, Donald E. Knuth, and Oren Patashnik. *Concrete mathematics*. Addison-Wesley Publishing Company, Advanced Book Program, Reading, MA, 1989. A foundation for computer science.

[Gla07] J. W. L. Glaisher. On the Numbers of Representations of a Number as a Sum of 2r Squares, Where 2r Does not Exceed Eighteen. *Proc. London Math. Soc. (2)*, 5:479–490, 1907.

[GM81] A. M. Garsia and S. C. Milne. A Rogers-Ramanujan bijection. *J. Combin. Theory Ser. A*, 31(3):289–339, 1981.

[GM93] David J. Grabiner and Peter Magyar. Random walks in Weyl chambers and the decomposition of tensor powers. *J. Algebraic Combin.*, 2(3):239–260, 1993.

[GNW79] Curtis Greene, Albert Nijenhuis, and Herbert S. Wilf. A probabilistic proof of a formula for the number of Young tableaux of a given shape. *Adv. in Math.*, 31(1):104–109, 1979.

[Gou71] H. W. Gould. *Research bibliography of two special number sequences.* Mathematica Monongaliae, No. 12. Department of Mathematics, West Virginia University, Morgantown, W. Va., 1971.

[GV85] Ira Gessel and Gérard Viennot. Binomial determinants, paths, and hook length formulae. *Adv. in Math.*, 58(3):300–321, 1985.

[Hag03] J. Haglund. Conjectured statistics for the q,t-Catalan numbers. *Adv. Math.*, 175(2):319–334, 2003.

[Hag04] J. Haglund. A proof of the q,t-Schröder conjecture. *Int. Math. Res. Not.*, (11):525–560, 2004.

[Hag08] James Haglund. *The q,t-Catalan numbers and the space of diagonal harmonics*, volume 41 of *University Lecture Series*. American Mathematical Society, Providence, RI, 2008. With an appendix on the combinatorics of Macdonald polynomials.

[Hai98] Mark D. Haiman. t,q-Catalan numbers and the Hilbert scheme. *Discrete Math.*, 193(1-3):201–224, 1998. Selected papers in honor of Adriano Garsia (Taormina, 1994).

[Han] Guo-Niu Han. Personal communications.

[HHL+05] J. Haglund, M. Haiman, N. Loehr, J. B. Remmel, and A. Ulyanov. A combinatorial formula for the character of the diagonal coinvariants. *Duke Math. J.*, 126(2):195–232, 2005.

[HL05] J. Haglund and N. Loehr. A conjectured combinatorial formula for the Hilbert series for diagonal harmonics. *Discrete Math.*, 298(1-3):189–204, 2005.

[HMZ12]   J. Haglund, J. Morse, and M. Zabrocki. A compositional shuffle conjecture specifying touch points of the Dyck path. *Canad. J. Math.*, 64(4):822–844, 2012.

[HP91]   Peter Hilton and Jean Pedersen. Catalan numbers, their generalization, and their uses. *Math. Intelligencer*, 13(2):64–75, 1991.

[HPW99]   Peter Hilton, Jean Pedersen, and Tamsen Whitehead. On paths on the integral lattice in the plane. *Far East J. Math. Sci. (FJMS)*, (Special Volume, Part I):1–23, 1999.

[HS19]   Han Hu and Chunwei Song. Generalized q-runyon numbers on truncated rectangle. *Util. Math.*, 112:265–285, 2019.

[HZZ16]   Han Hu, Feng Zhao, and Tongyuan Zhao. On S-(p, q)-Dyck paths. *Ars Combin.*, 125:225–246, 2016.

[Kan91]   Robert Kanigel. *The man who knew infinity*. Charles Scribner's Sons, New York, 1991. A life of the genius Ramanujan.

[Kas10]   Anisse Kasraoui. Ascents and descents in 01-fillings of moon polyominoes. *European J. Combin.*, 31(1):87–105, 2010.

[KdM13]   Joseph P. S. Kung and Anna de Mier. Catalan lattice paths with rook, bishop and spider steps. *J. Combin. Theory Ser. A*, 120(2):379–389, 2013.

[Kla03]   Martin Klazar. Bell numbers, their relatives, and algebraic differential equations. *J. Combin. Theory Ser. A*, 102(1):63–87, 2003.

[Knu73]   Donald E. Knuth. *The art of computer programming. Volume 3.* Addison-Wesley Publishing Co., Reading, Mass.-London-Don Mills, Ont., 1973. Sorting and searching, Addison-Wesley Series in Computer Science and Information Processing.

[KO92]   Ian Kiming and Jørn B. Olsson. Congruences like Ramanujan's for powers of the partition function. *Arch. Math. (Basel)*, 59(4):348–360, 1992.

[KP86]   Germain Kreweras and Yves Poupard. Subdivision des nombres de Narayana suivant deux paramètres supplémentaires. *European J. Combin.*, 7(2):141–149, 1986.

[KP10]   Markus Kuba and Alois Panholzer. On the area under lattice paths associated with triangular diminishing urn models. *Adv. in Appl. Math.*, 44(4):329–358, 2010.

[Kra89]   Christian Krattenthaler. Counting lattice paths with a linear boundary. II. q-ballot and q-Catalan numbers. *Österreich. Akad. Wiss. Math.-Natur. Kl. Sitzungsber. II*, 198(4-7):171–199, 1989.

[Kra95]   C. Krattenthaler. The major counting of nonintersecting lattice paths and generating functions for tableaux. *Mem. Amer. Math. Soc.*, 115(552):vi+109, 1995.

[Kra01]   C. Krattenthaler. Permutations with restricted patterns and Dyck paths. *Adv. in Appl. Math.*, 27(2-3):510–530, 2001. Special issue in honor of Dominique Foata's 65th birthday (Philadelphia, PA, 2000).

[Kra15]   Christian Krattenthaler. Lattice path enumeration. In *Handbook of enumerative combinatorics*, Discrete Math. Appl. (Boca Raton), pages 589–678. CRC Press, Boca Raton, FL, 2015.

[Kre66]   Germain Kreweras. Sur une extension du problème dit "de Simon Newcomb". *C. R. Acad. Sci. Paris Sér. A-B*, 263:A43–A45, 1966.

[Kre86]   Germain Kreweras. Joint distributions of three descriptive parameters of bridges. In *Combinatoire énumérative (Montreal, Que., 1985/Quebec, Que., 1985)*, volume 1234 of *Lecture Notes in Math.*, pages 177–191. Springer, Berlin, 1986.

[KS96]   Christian Krattenthaler and Robert A. Sulanke. Counting pairs of nonintersecting lattice paths with respect to weighted turns. In *Proceedings of the 5th Conference on Formal Power Series and Algebraic Combinatorics (Florence, 1993)*, volume 153, pages 189–198, 1996.

[KS99]   D. Kim and D. Stanton. Three statistics on lattice paths. In *Algebraic methods and q-special functions (Montréal, QC, 1996)*, volume 22 of *CRM Proc. Lecture Notes*, pages 201–214. Amer. Math. Soc., Providence, RI, 1999.

[KSY07]   Joseph P. S. Kung, Xinyu Sun, and Catherine Yan. Two-boundary lattice paths and parking functions. *Adv. in Appl. Math.*, 39(4):515–524, 2007.

[Lab93]   Jacques Labelle. On pairs of noncrossing generalized Dyck paths. *J. Statist. Plann. Inference*, 34(2):209–217, 1993.

[Las74]   Alain Lascoux. Polynômes symétriques et coefficients d'intersection de cycles de Schubert. *C. R. Acad. Sci. Paris Sér. A*, 279:201–204, 1974.

[Las75]   Alain Lascoux. Tableaux de Young et fonctions de Schur-Littlewood. In *Séminaire Delange-Pisot-Poitou, 16e année (1974/75), Théorie des nombres, Fasc. 1, Exp. No. 4*, page 7. 1975.

[Lot97]   M. Lothaire. *Combinatorics on words*. Cambridge Mathematical Library. Cambridge University Press, Cambridge, 1997. With a foreword by Roger Lyndon and a preface by Dominique Perrin, Corrected reprint of the 1983 original, with a new preface by Perrin.

[LSW11]   Dang-Zheng Liu, Chunwei Song, and Zheng-Dong Wang. On explicit probability densities associated with Fuss-Catalan numbers. *Proc. Amer. Math. Soc.*, 139(10):3735–3738, 2011.

[LY90]   Jacques Labelle and Yeong Nan Yeh. Generalized Dyck paths. *Discrete Math.*, 82(1):1–6, 1990.

[Mac60]   Percy A. MacMahon. *Combinatory analysis*. Two volumes (bound as one). Chelsea Publishing Co., New York, 1960.

[Mac78]   Percy Alexander MacMahon. *Collected papers. Vol. I*. MIT Press, Cambridge, Mass.-London, 1978. Combinatorics, Mathematicians of

Our Time, Edited and with a preface by George E. Andrews, With an introduction by Gian-Carlo Rota.

[Mac95]   I. G. Macdonald. *Symmetric functions and Hall polynomials.* Oxford Mathematical Monographs. The Clarendon Press Oxford University Press, New York, second edition, 1995. With contributions by A. Zelevinsky, Oxford Science Publications.

[Mac04]   Percy A. MacMahon. *Combinatory analysis. Vol. I, II (bound in one volume).* Dover Phoenix Editions. Dover Publications, Inc., Mineola, NY, 2004. Reprint of *An introduction to combinatory analysis* (1920) and *Combinatory analysis. Vol. I, II* (1915, 1916).

[Moh79]   Sri Gopal Mohanty. *Lattice path counting and applications.* Academic Press [Harcourt Brace Jovanovich Publishers], New York, 1979. Probability and Mathematical Statistics.

[MS08]   Toufik Mansour and Yidong Sun. Bell polynomials and k-generalized Dyck paths. *Discrete Appl. Math.*, 156(12):2279–2292, 2008.

[MSS12]   Toufik Mansour, Mark Shattuck, and Chunwei Song. q-analogs of identities involving harmonic numbers and binomial coefficients. *Appl. Appl. Math.*, 7(1):22–36, 2012.

[Nar55]   Tadepalli Venkata Narayana. Sur les treillis formés par les partitions d'un entier et leurs applications à la théorie des probabilités. *C. R. Acad. Sci. Paris*, 240:1188–1189, 1955.

[Nar59]   T. V. Narayana. A partial order and its applications to probability theory. *Sankhyā*, 21:91–98, 1959.

[Nar79]   T. V. Narayana. *Lattice path combinatorics with statistical applications*, volume 23 of *Mathematical Expositions*. University of Toronto Press, Toronto, Ont., 1979.

[Pec14]   Oliver Pechenik. Cyclic sieving of increasing tableaux and small Schröder paths. *J. Combin. Theory Ser. A*, 125:357–378, 2014.

[PR17]   Peter Paule and Cristian-Silviu Radu. A new witness identity for
           $11 \mid p(11n + 6)$. In *Analytic number theory, modular forms and
           q-hypergeometric series*, volume 221 of *Springer Proc. Math. Stat.*,
           pages 625–639. Springer, Cham, 2017.

[Pri97]   A. L. Price. *Packing densities of layered patterns*. PhD thesis, Univer-
           sity of Pennsylvania, 1997.

[Pro08]   Helmut Prodinger. Human proofs of identities by Osburn and Schnei-
           der. *Integers*, 8:A10, 8, 2008.

[Ram00]   S. Ramanujan. Some properties of $p(n)$, the number of partitions of
           n [Proc. Cambridge Philos. Soc. **19** (1919), 207–210]. In *Collected
           papers of Srinivasa Ramanujan*, pages 210–213. AMS Chelsea Publ.,
           Providence, RI, 2000.

[Ric15]   Thomas M. Richardson. The super Patalan numbers. *J. Integer Seq.*,
           18(3):Article 15.3.3, 8, 2015.

[Rio68]   John Riordan. *Combinatorial identities*. John Wiley & Sons Inc., New
           York, 1968.

[Rio69]   John Riordan. Ballots and trees. *J. Combinatorial Theory*, 6:408–411,
           1969.

[Ruk11]   Josef Rukavicka. On generalized Dyck paths. *Electron. J. Combin.*,
           18(1):Paper 40, 3, 2011.

[Sag91]   Bruce E. Sagan. *The symmetric group*. The Wadsworth & Brooks/Cole
           Mathematics Series. Wadsworth & Brooks/Cole Advanced Books &
           Software, Pacific Grove, CA, 1991. Representations, combinatorial
           algorithms, and symmetric functions.

[San16]   Carlo Sanna. On the p-adic valuation of harmonic numbers. *J. Number
           Theory*, 166:41–46, 2016.

[Sat89]   Masako Sato. Generating functions for the number of lattice paths
           between two parallel lines with a rational incline. *Math. Japon.*,
           34(1):123–137, 1989.

[Sch70]   E. Schröder.   Vier combinatorische Probleme.   *Zeit. f. Math. Phys.*, 15:361–376, 1870.

[Slo]   N. J. A. Sloane. The On-Line Encyclopedia of Integer Sequences. http://www.research.att.com/~njas/sequences.

[Slo73]   N. J. A. Sloane. *A handbook of integer sequences.* Academic Press [A subsidiary of Harcourt Brace Jovanovich, Publishers], New York-London, 1973.

[Slo18]   Neil J. A. Sloane. The on-line encyclopedia of integer sequences. *Notices Amer. Math. Soc.*, 65(9):1062–1074, 2018.

[Son05a] Chunwei Song. The generalized Schröder theory. *Electron. J. Combin.*, 12:Research Paper 53, 10 pp. (electronic), 2005.

[Son05b] Chunwei Song. On permutation paths and signed permutation paths. *Far East J. Math. Sci. (FJMS)*, 17(3):281–298, 2005.

[Sta]   Richard P. Stanley. Bijective proof problems. Available at https://math.mit.edu/~rstan/bij.pdf.

[Sta71]   Richard P. Stanley. Theory and application of plane partitions. I, II. *Studies in Appl. Math.*, 50:167–188; ibid. 50 (1971), 259–279, 1971.

[Sta97]   Richard P. Stanley. *Enumerative combinatorics. Vol. 1*, volume 49 of *Cambridge Studies in Advanced Mathematics*. Cambridge University Press, Cambridge, 1997. With a foreword by Gian-Carlo Rota, Corrected reprint of the 1986 original.

[Sta99]   Richard P. Stanley. *Enumerative combinatorics. Vol. 2*, volume 62 of *Cambridge Studies in Advanced Mathematics*. Cambridge University Press, Cambridge, 1999. With a foreword by Gian-Carlo Rota and appendix 1 by Sergey Fomin.

[Sta15]   Richard P. Stanley. *Catalan numbers*. Cambridge University Press, New York, 2015.

[Sul82]   Robert A. Sulanke. q-counting n-dimensional lattice paths. *J. Combin. Theory Ser. A*, 33(2):135–146, 1982.

[Sul00]   Robert A. Sulanke. Counting lattice paths by Narayana polynomials. *Electron. J. Combin.*, 7:Research Paper 40, 9 pp. (electronic), 2000.

[Sul03]   Robert A. Sulanke. Objects counted by the central Delannoy numbers. *J. Integer Seq.*, 6(1):Article 03.1.5, 19 pp. (electronic), 2003.

[Sul05]   Robert A. Sulanke. Three dimensional Narayana and Schröder numbers. *Theoret. Comput. Sci.*, 346(2-3):455–468, 2005.

[SW02]   Zvezdelina Stankova and Julian West. A new class of Wilf-equivalent permutations. *J. Algebraic Combin.*, 15(3):271–290, 2002.

[SY12]   Chunwei Song and Catherine Yan. Descents of permutations in a Ferrers board. *Electron. J. Combin.*, 19(1):Paper 7, 17, 2012.

[Tch52]   P. Tchebycheff. Mémoire sur les nombres premiers. *J. de mathématiques pures et appliquées*, Sér. 1:366–390, 1852.

[vLW92]   J. H. van Lint and R. M. Wilson. *A course in combinatorics*. Cambridge University Press, Cambridge, 1992.

[Wil94]   Herbert S. Wilf. *generatingfunctionology*. Academic Press Inc., Boston, MA, second edition, 1994.

[Win69]   Lasse Winquist. An elementary proof of p(11m+6) = 0 (mod 11). *J. Combinatorial Theory*, 6:56–59, 1969.

[Wor83]   J. Worpitzky. Studien über die bernoullischen and eulerschen zahlen. *Journal für die reine und angewandte Mathematik*, 94:203–232, 1883.

[WW13]   Robin Wilson and John J. Watkins, editors. *Combinatorics: ancient and modern*. Oxford University Press, Oxford, 2013.

[Zha15]   Tongyuan Zhao. *The enumeration theory of S-(p, q)-lattice paths*. PhD thesis, Peking University, 2015.

Index

Printed in the United States
by Baker & Taylor Publisher Services